TEN THOUGHTS
ABOUT
TIME

BODIL JÖNSSON

Robinson
London

Constable & Robinson Ltd
3 The Lanchesters
162 Fulham Palace Road
London W6 9ER
www.constablerobinson.com

First published in Sweden as *Tio Tankar om Tid* by
Brombergs Förlags, 1999

First published in the UK by Constable,
an imprint of Constable & Robinson Ltd 2003

This paperback edition published by Robinson,
an imprint of Constable & Robinson Ltd 2005

A copy of the British Library Cataloguing in
Publication Data is available
from the British Library

ISBN 1-84529-050-X (pbk)
ISBN 1-84119-543-X (hbk)

Printed and bound in the EU

1 3 5 7 9 10 8 6 4 2

CONTENTS

PREFACE

I have been thinking about time, the subject of this book, on and off for the last twenty years or so, and my thoughts have already been aired in letters and conversations, articles, lectures, radio broadcasts and books.* The present book was written on request. It may sound a bit conceited, but there has been a strong demand for it. This is not surprising, as I have come to realize that I am far from alone in my preoccupation with time. Other people with similar concerns have wanted to

---

* E.g. in one of the *Summer* programmes on Swedish Radio; in the (audio-) book *Den annorlunda plånboken* [*An Alternative Wallet*] (1997) written together with K Örnfjäll.

talk, exchange views and develop their ideas. Thinking aloud even gives us the chance to take pleasure in the passage of time.

Time is not an issue that can be dealt with once and for all. I am sure most people think of it in a way not unlike my own image: as a staircase with four steps which I tramp up and down all my life.

To get on to Step One, you must be able to think of time without being depressed about it. You have to stop the cycle of 'Damn, I'm always so short of time,' 'No, I haven't got time' and 'I've no hope of getting everything done in time.' Once you achieve that, it is possible to develop new relationships with time.

A methodical system for organizing your time and how you use it will help you clamber on to Step Two. From this vantage point it becomes easier to link and structure your ideas about time, and even smile forgivingly at yourself as you find yourself starting to worry away at your favourite subject again.

One more step up, and you become more detached still: by now you can articulate your views on time and your life as part of its steady flow. This is important, because it enables you to express yourself, both in your own mind and to other people. It is only when you can see your thoughts as a coherent whole that you can develop them further. And then you don't have to start from the beginning each time!

Standing, as it were, on your own shoulders, you are now in a position to crawl up on to Step Four. Here you can contemplate your ideas, and evaluate your own ways of thought in comparison with those of others.

Up there on Step Four you may feel you can relax, believing that you now have a clear idea of what matters to you in your relationship with time. And then life takes an unexpected turn or – all too often! – you lapse into your old ways, and the stoic contemplation comes to an abrupt end. Back you go to one of the lower steps. Still, it gets easier to climb up each time.

I will not mention my staircase again; it is just a metaphor I devised for my own use. However, I will come back to experiences and thoughts I have had while climbing up and down its steps. I hope you will find some of what I say of interest to you. For me, writing this has been a sheer pleasure. Sorting out past ideas, finding new ones and rearranging them in different patterns has been as enjoyable as picking a bunch of flowers.

The date of the book, as written out below, is an

added bonus, a cherry on the cake: three 1s and three 9s is a very special sequence! Almost as special as a 2 and three 0s.

Bodil Jönsson
Near Stenshuvud, 1.1. 1999

I

TIME – YOUR BEST ASSET

I don't hero-worship people as a rule, but Granny, my father's mother, is an exception. She died before my seventh birthday, but looms large in the few distinct memories from my childhood. Granny was important to me for many reasons, of course, but here I have one particular quality of hers in mind – the way she always found time for what mattered to her.

By our standards she lived in straitened circumstances, often short of food, warmth and light. Yet she never seemed to run out of time – though this was not how she saw it. She simply did not think of life in terms of time. It has taken just two generations to create my own time-dependency and, more generally, a society whose members all feel time-deprived. It is an odd deprivation, for time is the one asset we all have a share in.

In Western Europe, a human life lasts on average 30,000 days and nights. This is our capital, an asset allocated to each individual. This makes it seem unreasonable, even inhuman, that time should have become classified as rationed goods. How did it come about – after half-a-century of steadily improving standards of living – that we moved from the calm pace of living in the 1950s to one that feels hectic and erratic rather than rhythmical? Perhaps the most important reason is the flexibility of the human mind. Our built-in mental rhythms are adaptable to a degree that can be counterproductive. Human creativity increasingly has to adjust to the demands of machinery, and so people become less able to control their timekeeping. In other words, human inventiveness, sensitivity and flexibility are in conflict with technology's predictability, lack of imagination and resistance to change. Or to put it another way, human beings, who are forgetful, illogical, disorganized and emotional, are trying to co-exist with technical devices that have excellent memories and are precise, logical, highly organized and reliable, but not adaptable. Guess which has to give way.

In evolutionary terms, this is a new problem for humans. Until recently, no one had made any connection between timekeeping and human characteristics. Leonardo da Vinci's famous image of man as the measure of all things is based on *geometry*. He felt no need, clearly, to consider anything more than a three-dimensional universe made up of lines, surfaces, volumes and their relative relationships. Nowadays we feel the need for a fourth dimension, a new image of man as the measure of time, that no one has yet invented.

A WALLET WITH A DIFFERENCE

What asset do you own that can be turned into either money or companionship? An asset that can also help you interact with your surroundings – whether natural or built environments, technology or goods in general – and help you develop your knowledge and your emotions?

That asset is your time.

Imagine a wallet with four compartments, only one of which contains money. Another is reserved for the people around you, both family and friends and more casual acquaintances. A third holds your non-human environment, made up of nature, buildings, things and your creative activities. The fourth and final compartment is for your inner self, your thoughts and feelings.

It might seem a dull, even pointless, exercise to subdivide time in this way. Swapping 'currencies' between the different compartments looks impossible in practice: you can't normally barter money for knowledge, or gadgets for human company. But time exists outside the wallet, like a gold standard. Time provides the one way of making exchanges between the currencies in your personal wallet. This means that time is your most essential resource.

It is instructive to try to separate your life into compartments, even though you don't normally deal with it like that. Most noticeable is the shockingly large proportion of effort, both your own and other people's, spent on filling the money compartment. What would be the effect of redistributing that effort? Above all, what would be the result of being more concerned with time, the main basis for exchange, the currency for investments and withdrawals? What if we decided to focus on our time, the gold standard itself, rather than the contents of any one compartment? Perhaps we would start planning seriously for a Time Protection Agency, along the lines of an Environmental Protection Agency.

THE MISSING KEYSTONE

Now is the time to rethink. Economic engines are running out of control and ecological systems are collapsing, as land, sea and atmosphere are asked to cope with horrendous degradation. No one seems able to cope with more than a limited range of these symptoms. Blinkered, we might be missing the crucial point – the keystone supporting a whole structure of interlocking events.

Remember Three Mile Island, site of a nuclear reactor incident which could have gone on to meltdown? The reason the reactor went critical was that everyone failed to grasp the real cause of its malfunction. The warning signals were there, and every single correction was carried out by the book, but they all miscarried because the operators misinterpreted the chain of cause and effect. The key event was a valve getting stuck. Nobody understood this, and so they made mistakes. Over and over again.

Maybe we will go on making mistakes about the way we manage our environment, social relationships and peace of mind, if we similarly fail to appreciate that our relationship with time is the keystone in the structure of our lives. What if this is the real cause of the mistakes we make? For instance, might not a different approach to experiencing time have more environmentally-friendly effects than piecemeal attempts at specific environmental improvements?

In any case, it must be a good thing to get away from the belief that time is money. It is always worthwhile to look for new ideas leading away from the notion that *cash* is life's gold standard. The dangers of simple greed may have been more obvious in the past than they are now. One Swedish example is the case made by the wives of agricultural workers against the demands of the men's union for cash payments in return for their labour – that is, for their time. Until then, payments had been in kind, and this system had been the core of family survival. The only money in the household used to be that paid to the women for the bits and pieces of needlework and laundry they managed to do on top of their many other tasks. Even though the cash in hand was worth very little – pin money – it mattered to the women. They felt threatened by the idea that all work, men's as well as women's, might come to be evaluated in cash only.

Much has happened since then, and saving time has become a growth industry in the industrially-developed nations. I can't resist giving you an example of how crazy the consequences can be. Assume that your work is 50 kilometres away from your home and that you commute by car. That means driving 100 kilometres every day, which we assume will take you about one hour. Let us calculate the cost, and although the actual figures are out of date – in some places they might seem positively utopian – the equation still works. Say it costs £20 to drive 100 kilometres and you are paid an average of £5 per hour (after tax) for your work. This means that it will take 4 hours of work to fund your driving. So the time involved is not 1 hour but 1 + 4 hours = 5 hours. At an average speed of 20 kilometres per hour, (100 kilometres over 5 hours) you might as well cycle all the way!

TIME TO RETHINK

Can the striking example above be made to apply in real life? It is no easy matter to persuade your employer that 'It makes sense for me to restructure my work schedule, so instead of working eight hours and driving for one every day, I will work for five hours and cycle for four.'

Individual decisions cannot be made in a vacuum. Our lives are intertwined with the economy, employment conditions, retail markets and the whole public sector, notably health care. Almost all of these require us to drive a car at some stage. But this doesn't mean that the quiet resolution 'I'd just as soon cycle' is wrong or pointless. Small, personal changes are easy enough to bring about. Once in a while, they might start a great revolution. The first push in the right direction can come from metaphors, calculations and comparisons that make us laugh at our superficially rational lifestyle.

I would like to get round to looking at as many areas as possible in just that way. Consider, for instance, the price of a standard plane ticket in the context of cycling! Could it be that we're in such a hurry being in a hurry that we have no time for anything else? I think so – often, at any rate. Things don't happen nearly as quickly as we believe. So-called time-gains tend to be paid for by the extra work we put in to make them.

When I come across someone explaining a new time-saving gizmo, I try to find the courage to ask 'And how are you going to use the time?' It's a good question, but also dangerous. It is meant to remind people of the keystone in the structure of their lives. Here you are, and time seems to be slipping out of your hands faster and faster. What can you do? You buy a time-saving device. But still time accelerates away from you, forcing you to find the money for more time-saving ideas…

I started thinking about these things in my early thirties. I had three young children and an exciting job. Every day, time rushed past more quickly than before. I was talking to a woman who seemed almost elderly to me then, but must have been in her early fifties. When I rather hesitantly tried to explain how anxious it made me to feel that I was losing my grip on the flow of events, she simply replied, 'If that's how you feel at your age, just wait till you're older!' I've never forgotten that – it went straight home!

There is one thing every scientist must learn to do: to think logically. If my older friend's observation was generally true, then the acceleration of time passing, which I had just begun to notice, would never decrease. Thus it followed that life would end much sooner than I wanted it to, because I enjoyed living very much.

TAKING TIME OUT

The seed had been sown. Quietly, I began a private project to 'stop time'. I tested what turned out to be an effective approach: rather than dashing about like crazed rat all the time, I would do nothing for a set period. Well, perhaps not exactly nothing. Being the person I am meant that there was always something going on.

This is how I did it: suddenly, I decided to take time off work, for almost two months after Christmas. The expression 'time out' hadn't yet reached Sweden, but that was what I wanted to achieve. I was neither ill, nor depressed, nor a burnt-out case: I just wanted to stop time.

During the first week I was busy tidying the attic and sanding floors, but I calmed down later. It was important to stay at home. My idea was to wait and watch, not travel or look for other things to do. Slowly, a sense of time stretching into eternity came back. The panic-stricken chasing about – 'what do I need to do next, what have I forgotten?' – and the neurotic anxiety – 'but it's pointless, pointless, what does it matter anyway?' – all gradually faded away.

Looking back, I think that time never again seemed to move as fast as it did before my time out. The trick could work more than once. If I ever feel like a hamster in a wheel again, I will repeat the exercise, and expect to return to normal life in better harmony with the wave-motion of time. I still take time out on a small scale. Of course it doesn't mean that my life always runs smoothly. It's just that when things get snarled up, I know how good it is to withdraw and have a quiet breather before starting out again. I'm still running in the wheel, but it makes a terrific difference to know it won't turn for ever – 'Now or sometime, I'll get out of this!'

Deep inside, I know I've got plenty of time. This can be annoying, as many inquiries from people who know me well, and even complete strangers, have taught me. The questions focus anxiously on *my* spare time, though unmistakably they are asking themselves about their own relationship to time. How is it possible to have plenty of time? How is it done?

NO ONCE-AND-FOR-ALL ANSWER

In a moment I'll get round to outlining a couple of smart time-controlling ideas for you to try. First, it is important to issue a warning. Your relationship with time is a very personal and variable one. It cannot be settled on the spot by any of the instant fixes that keep turning up in the context of time-management. When you have taken on board the need to change it, you must also accept the need to scrutinize the workings of your own mind, thoroughly and often. It is to stimulate your sense of personal involvement that I write this book in the same spirit as Eyvind Johnson wrote his brilliant novels about the philosopher Krilon – they are based on his analysis of the importance of conversation, internal as well as external.*

I like the philosophers' approach to debate. The arguments are taken for walks, as it were, but made to return to certain points from different angles and at different times. This is how one learns, after all. Learning is not about once-and-for-all answers or exact repetition, but finding out about the variations that may or may not lead to the same result. Also, though it is impossible to deal with anybody's relationship to time once and for all, it can be crucial to learn to recognize the symptoms of it foundering. When a time-vortex threatens, it's useful to have a trick or two up your sleeve. The best thing is to be able to laugh at yourself. Laughter always helps.

---

* Eyvind Johnson (1900–1976), Swedish writer and co-recipient of the 1974 Nobel Prize for Literature, is the author of more than 40 novels and short-story collections, including the Krilon trilogy, written between 1942–1944 (*The Krilon Circle*, *Krilon's Journey* and *Krilon Himself*).

FREE TIME

It is as patently untrue to claim always to have plenty of time as it is to insist on always being short of time. Either way, after becoming conscious of time in a new way, you will find yourself changing your priorities, with regard to both activities and time allocations. If your diary is packed with entries and reminders, you might become more aware of the way they tend to cancel each other out. This can only be done by prioritizing and re-prioritizing. If you want to liberate yourself from the tyranny of time altogether, then again prioritization must be part of the answer. You have to be selective in order to do what you really want. Select, so that you have time to treat yourself. Reorganize your life so that the option of free time becomes available. Make room for innovation in what you think and what you do. Such changes need time, space and considerate people around you. Above all, you must stop distracting yourself by allowing minor interventions to dominate your life.

Keeping a lot of balls in the air should not mean that you have to juggle with them all simultaneously. Peaceful moments are necessary to allow you to catch the balls one by one. By all means throw them into the air in any order that seems right at the time, but never let them become too many. How many is too many is for each individual to decide. Personally, I can manage a fair number, but I know when yet one more would suddenly mean overload. If that happens, I lose control altogether, and usually drop even the last ball I have tried desperately to hang on to. The first sign that I'm trying too hard is when the interval between thought and action lengthens, as I add yet more balls to my juggling act.

IN PRAISE OF DIVERSITY

There was a time when I assumed that growing older would be like slipping down a funnel-shaped, swiftly narrowing passage: life would become ever more restricted and uniform. From my experience to date, I can only conclude that I got it utterly wrong. Nobody told me this when I was young, so now I feel duty-bound to let everyone know – the excitement of being alive need not diminish with age. On the contrary, experiences are often more vivid the older one gets. Part of the reason is presumably that the more you have already been through, the more relevant associations to new events you have accumulated. I know that thirty years ago I was simply unable to take in many things – just as grass to most people is green plant-stuff, but to the botanist it is hundreds of distinct species. By taking a long-term interest in music, the listener learns to shift between enjoying the whole and distinguishing between the voices of single instruments. Subtle comparisons between different compositions become possible, which in turn enriches the entire experience. The spiral of enjoyment and learning is twisting the right way – unless boredom sets in, of course.

Like many other older people, I find that a nucleus of self-confidence has formed inside me. I dare to act, to feel, to test and to experience. This means that my mental territory is expanding as much as, if not more than, when I was younger. I feel that the flow through the funnel is the reverse of what I first thought, towards a wider space.

Sometimes I'm struck by how little I remember from the periods when my life was running on rails. When all went well, what remains is no more than a sense of having been happy. That's a good thing in itself, but also somewhat disquieting. My strongest memories are of difficult episodes. Not that I keep going over past problems, and as for grudges, I hardly remember them at all: blissfully, I seem unable to become bitter. What I value is the sense of having endured and coped with something difficult. This strengthens my willingness to try new things, including a new relationship with time.

INTERLEAVING ACTIVITIES

There are both advantages and disadvantages in trying to interleave activities and add intervals of doing nothing. The basic prerequisite is that you have to be able to cope with the non-productive, often irritating stress caused by shifting from one activity to another.

- An advantage: it's part of the paradigm 'Thinking Takes Time' (TTT; cf. Ch. 5) that it's sometimes necessary to be bored with a particular chain of thought, i.e. allow it to mature under the influence of those subterranean mental processes over which we have so little control. It is impossible constantly to think new thoughts. Repetition is essential. While this is going on, it can be useful to take a break or to do something different.

- A disadvantage: interleaving activities can lead to

disproportionately long pauses, as you exit one mindset and enter another. It seems to take too long to come to a halt and then start up again for anything useful to be done. I shall write more about this problem under the heading Set-up Time (Ch. 3).

- An advantage: you escape monotony.

- A disadvantage: too much diversity can lead to the wrong things being given priority, for instance preferring easy, quick activities with short-term rewards to the harder ones, which might be both more fun and more useful in the long run.

STANDING STILL AND MOVING ON

Stillness and movement enhance different characteristics. This is true for both people and matter. Consider a jar of syrup: how much can you find out about the syrup without tilting the jar? Note the colour, sense the taste perhaps. But syrup is semi-solid, and it is hard to find out anything about its fluidity until you have practically turned the jar on its side. Only then can its dynamic properties be observed.

Dynamic properties are never revealed in a static state. It's like discovering a new personality altogether, when you encounter dynamic qualities in someone previously only known to you in quiet settings. The person sharing the warmth of your campfire in the wilds is someone quite different from the colleague in the busy routine of your city life. No wonder the qualities listed in recruitment advertisements are different nowadays. In the past they focused on static qualities such as reliability and persistence, but now the emphasis is on creativity, sensitivity and initiative.

THE PRESENT

In the past, time was sovereign in nature. Its rule was a wonderful way of preventing disorderly events. Nowadays, it is as though the ordering function of time has been cancelled out by demands that almost everything should happen simultaneously. The invisible pressure, which used to sort the temporary from the lasting has lost its effectiveness. Once this pressure squashed fashionable but weak inventions and research. Our attempts to speculate about the distant future seem irrelevant now, because the distance in fact seems so short.

Is the nearness of the future frightening? Well yes, perhaps it is. But then, it's amazing to be alive *just now*, when every possible experience is open to you instantly. Your time is not just your allotted time, but the moment you have been given to exploit *the present*.

The present is an important concept. True, you don't know all the people you would like to meet, but in principle there is nothing to stop you from arranging to talk to every living human being you fancy.

It is, of course, impossible to talk to people who are no longer alive or not yet born. The consequences of contemporaneity can also be seen in the relationships between states. If a country challenges another by, for instance, threatening to reduce its neighbour's water supply, the conflict requires immediate solutions. When mismanagement by anyone, anywhere, endangers the water supply of future generations, the act has a lack of immediacy that usually means postponing any attack on the source of the trouble. No talks can take place between the opposing sides – those who cause the trouble and their victims.

SIGNS TO TELL YOU THAT
TIME IS YOUR BEST ASSET

The importance of time is at its most obvious at births and deaths. The small, new being has all its time ahead of it. The child will also annex large parts of your time. When someone has died, sharing your time with the dead person is no longer possible. At such moments of realization, thoughts about your own allocation of some 30,000 days and nights can become both sombre and very personal. Do we even dare approach the core questions? To be rootless is not just a psychological or social characteristic: there is a kind of rootlessness in the present too. But thoughts such as these are practically taboo.

Our everyday life is filled with events, people and places. Now and then we claim to live for the present, or at least want to live only in the present. But unless you know what you are likely to do next month, your grip on the present becomes weak to the point of paralysis. A modicum of expectation is essential for living in the present. The same applies to the past and our access to memories. Then, now and later are pivotal concepts to movements within life.

My aim in this book is to make you feel more

relaxed about time but also more aware of its special functions. I want to draw your attention to time as a unique dimension, even in the most everyday of events. 'Time – it's your best asset' is a fact, but it can be sensed as a source of joy or a stimulus to thought or an exhortation or a challenge – it's up to you.

2

Clock-time and Lived-time

The mind has got built-in sensors that adjust to circumstances. When you have just been told that you are going to have a baby, pregnant women and parents pushing prams seem to be everywhere. Once you have got it into your head that the mole on your back might be malignant, it seems to itch all the time. Hunting for mushrooms easily becomes spotting thousands of mushroom-coloured leaves.

To a great extent, our sensors direct what we learn. Maybe we should reset them to act as detectors rather than receivers.

Knowledge can play a role in the mental scanning process, as well as in determining the type and intensity of the experience. Compared with the amateur, the astronomer observes the night sky in a different way and in more vivid detail. The same is true of, say, the biologist confronting a diversity of species, or the music critic listening to a concert.

HOW MANY 'SOONS' MAKE UP
A QUARTER OF AN HOUR?

Time is a special case, though. We have few sensors for detecting time, and cannot become experts in experiencing it. There are many ways to understand it better, like learning to keep time or training to become, say, a watchmaker or a logistician, whose job is to co-ordinate timed processes. Some people can lead projects because they have the ability to adjust the time available to the tasks in hand. Others become physicists with a taste for research into the experimental and theoretical aspects of time. True, physics is no use when it comes to mysteries like 'How long is quarter of an hour?' or 'How many "soons" make up "a while"?' Physicists are better than others when it comes to defining time units, but it does not help me when I try to work out the meaning of 'a while'.

I believe that we should live by two kinds of time,

and keep the distinctions clear: *clock-time*, based on atomic periodicity, as opposed to *lived-* or *personal time*. There are actually very few similarities between them. Human beings are poor at measuring objective clock-time. We have internal (biological) clocks, but they change their settings from day to day, from hour to hour, even from minute to minute – they are technically unreliable, in other words.

But there is no question that lived-time is as real as mechanistic clock-time, although in another dimension. On the one hand, I have a sense of time that is mine alone, and on the other, in order to keep appointments, I need an artificial but shared standard of measurement. Odd, isn't it? Or is it?

Actually, it's not odd at all. Technical artifice is often created precisely to bridge the gaps between people. There are those who feel that technology is inhuman, but that isn't right. A human being can be inhuman, but a technical device can't. Though it can be badly constructed – a device depends on technology, and when it works poorly, it fails to be good enough technically. Communication devices are the most common interfaces between people and technology. The telephone was invented because we defined the wish to call other people. Railways were built because we had people to visit. Clocks would not have been devised if human encounters did not demand agreed timekeeping.

DEFINED, MEASURABLE TIME: CLOCK-TIME

Artificial clock-time is the easier of the two to describe, because people agree about it. Measurable time, like other measurable physical phenomena, has been defined, together with units of measurement. In a book I co-authored called *Experimental Physics*, we wrote that the duration of a second makes sense, just as it makes sense for a kilogram to weigh what it weighs and for a metre to measure the distance of a metre.* These units fit human dimensions. One second, the basic unit for measuring time, is close to the time between two heartbeats. The kilogram unit used to measure mass is appropriate because a human being weighs a multiple of kilograms. One metre – well, we're between one and two metres tall, generally. The temperature unit one degree Celsius is good, because a few degrees

---

* Co-authored with Nina Reistad.

Celsius are about as much as our temperature sensors can distinguish between.

There have been problems with getting these units universally accepted, but in retrospect it has worked out well. Now people in all kinds of places accept the units of 1 metre, 1 kilogram and 1 degree Celsius, and can quite easily convert between their own units and the standard ones. Remarkably, in the case of time it turned out that one second is the accepted unit practically everywhere.

DISTANCE AND TIME

Units with stable values are necessary for accurate measurements. At first people put their faith in observable phenomena that appeared to stay unchanged over time. The shapes and movements of astronomical bodies, and the laws that governed them, seemed to provide stability. For instance, the base unit for measuring distance, one metre, became defined as exactly one ten-millionth of the distance between the equator and the North Pole. The dimension of time, and the units for measuring it, became tied to astronomical 'constants'.

However, when measurements improved through better technology, it was soon discovered that celestial phenomena are nowhere near as unchanging as had been assumed. As James Clerk Maxwell stated: 'If we require absolute, unchanging units with which to describe distance, time and mass, we must not make these dependent on the movements or masses of the planets, but confine ourselves to the wavelengths, frequencies and masses associated with stable, unchanging and totally identical atoms.' *

The result was an increasingly swift development of new measuring techniques, and so new units. The definition of one second was based on atomic radiation frequencies, and the definition of one metre recently became directly linked to that of the second: for the curious, some details are given in a footnote below.†

However, I know no one who is able to make these definitions part of their own way of experiencing life. The strength and importance of the new units lies

---

* Scottish mathematician and physicist (1831–1879); e.g. from *The Scientific Letters and Papers of James Clerk Maxwell: 1862-1873* by James Clerk Maxwell, P.M. Harman (Editor).

† Since 1967, a second has been defined as exactly 9,192,631,770 oscillations or cycles of the caesium (Cs^{133}) atom's resonant frequency. Since 1983, a metre is defined as the distance travelled by light in a vacuum during $1/299,792,458^{th}$ of a second.

precisely in their objectivity, as independent as possible of human beings. One interesting general point is that distance is no longer defined as a separate entity, but is linked to time. The reason is that in the 1970s it became possible to measure the speed of light in a vacuum, expressed as metres per second, more accurately than the distance of 1 metre, and the metre was duly re-defined. Clock-time is now the predominant measure, which fits well with how we experience our world: in terms not of distance but of time.

PERSONAL, LIVED-TIME

Physical phenomena are designed to be untouchable. Humankind has made it a matter of honour to find constants beyond human manipulation, first in the cosmos and later in the micro-cosmos of the atomic structure. But you might prefer an ordinary, mechanical clock, like the writer of the verse:

> I'd hate to use a digital miracle
> to measure my time going by.
> Instead my small pocket oracle
> lets me dream of immortality.

Your personal time is yours. You can manage it as you like, but also invest it with thoughts and feelings of your own. We tend to want our personal time to last, perhaps for an eternity. Meanwhile, we keep chasing clock-time and try to use it effectively. We chop it up into tiny bits to get flexibility. Then we buy gadgets to help us save time. Most of these interventions are counter-productive, if it is plenty of lived-time that we are really looking for.

Sometimes one man's time-saving becomes another's waste of time. One example is the way mobile phones allow other people's time anxieties to invade the rail traveller's 'free' stretch of time between stations. There is an argument for keeping mobile phone addicts in separate compartments, like smokers, as some train companies now do.

TIMELY THOUGHTS

While I was busy inventing my clock-time and lived-time, the rest of the world has not ignored the question of our relationship to time. It is a virtual certainty that if you are pondering a particular issue, people elsewhere on our planet are doing the same thing. It's part of the

Zeitgeist – if you like, some lines of thought are part-icularly timely.

Even so, I was surprised to find, when browsing through a newspaper, that the German philosopher Peter Heintel had independently arrived at many of the ideas about time that I had felt were my own. In 1990, Heintel founded a society called Tempus, whose task was to extend – or delay, or slow down – time.* Even though there is much humour in its proceedings, Tempus is a fundamentally serious organization. It organizes formal symposia, but members are also encour-aged to present practical examples of their own and other people's ideas on personal time management. The society publishes and sells written and filmed material. One of its books, called *Signs of Time (Zeitzeichen)*, is marketed as the gift for people who have everything – except time.

I have read a great deal of their publications, and was initially amazed to find that concepts like 'Chronos'

---

* 'Verein zur Verzögerung der Zeit'. Tempus has about 1,000 members of branches in Germany, Austria, Italy, Switzerland and Sweden. It has organized annual international symposia on aspects of time since 1991.

and 'Kairos' were exact replicas of my 'clock-time' and 'lived-time' respectively. Here are some quotes from texts published by Tempus: 'Live faster and life ends more quickly,' 'I have time, therefore I am' and 'Olives ripen no faster because you inspect them more often.'

Is anyone surprised at this coincidence of thought? We live at the same time and discuss contemporary concerns, far removed from the old, daily battles for survival. It is only now that we have the time to think systematically about time. We have to, if we want to get out of the fixation on clock-time that industrialization has brought in the wake of its conviction that 'time is money'.

HOW TO GET MORE LIVED-TIME

It seems likely that many people will read this book in search of answers to the question 'How do I personally get more lived-time?' Here is my first – but not my last – answer: try to become conscious of lived-time and realize that what the clock says is not all-important. There are other ways of measuring time.

Once you feel certain in yourself that you have a personal sense of time, you can start working on how to get more out of it. You will find that once you have learnt to manage your 'Set-up time' (Ch. 3) sensibly and consciously, new options for using your lived-time will open up. This will also happen when you find you can balance undivided and sub-divided time (Ch. 4) successfully.

3

SET-UP TIME

*C*oncentrating is more difficult if there are potential disturbances about. It's quite easy to disturb yourself. To test this, try a simple series of subtractions – 7 from 478, then 14 from 471, 28 from 443, etc. – and carry on while you read this page.

See what I mean? What and how much you can do in parallel varies, but it is limited. Some people must stop walking to chew gum. Cycling and running are fairly automatic activities, but I always lose speed when I start thinking hard about something. Women are supposed to be able to keep many balls in the air simultaneously. It's true enough, I'm sure, and I know many men who are very good at it too. But it depends on the balls and how many of them there are: there are limits.

By the way, how are you getting on with your sub-tractions? Have you even reached a number below 400? Did you notice that I had made a mistake (471 − 14 = 457, not 443)? You agree that we are not made for simultaneous tasks of this kind, don't you?

There are tasks better carried out not only in circumstances that allow you to concentrate, but also after allowing for a certain 'set-up time'. The set-up time is the time it takes to get things ready − to prepare for doing something. When men and horses worked the forests together, the logger's set-up time was spent harnessing the horse to the timber-sledge. In workshops the set-up time is the time it takes to re-jig the machines, and in a restaurant kitchen, it is the time set aside by the chef for his 'mise-en-place', i.e. arranging his essential equipment before starting to cook in earnest. Such preparation time is part not only of long-established crafts, but also of modern project planning.

This chapter is about recognizing that different set-up times are required for different tasks. I shall spend some time discussing how to profit most from invest-ments in set-up time. All time is precious, and set-up time must not be wasted. The forester would not take his horse and sledge into the winter forests without bringing back a load of logs. The chef uses his 'mise-en-place' to get food ready to serve. If the set-up time is spent in concentrated thought, that too should be put to good use. Take care of it and don't squander it by allowing yourself to be disturbed by telephone calls and the like.

DIFFERENT TASKS – DIFFERENT SET-UP TIMES

It sometimes helps to structure tasks according to type. Look at the grid of categories below. What do you think of first? Maybe it's not of the set-up times.

EASY & FUN	EASY & BORING
HARD & FUN	HARD & BORING

But the set-up times in the different categories are different, perhaps more crucially than you may have thought. When tasks of all four types are waiting, it's tempting to start with the Easy ones. Usually, the set-up time for Easy tasks is practically nil. Most of us feel a need to deal with Easy & Boring jobs first in order to clear our minds for the rest.

This would be fine if personal time were unlimited,

but as it isn't, the risk is that you may never tackle the Hard tasks. Usually, it's not the case that Hard jobs are unimportant or impossible. Far from it. The problem tends to be the apparently unproductive set-up time that must precede difficult undertakings.

In other words, I believe the problem is less about performance anxiety than about anxiety about the preparation time. However, completing one big thing will bring rewards which in the end are likely to be much more satisfying than those of many small tasks. The challenge is to make a conscious decision to prioritize something hard and work through the set-up time of the task. The other snag is the problem of making people around you accept that, whatever their initial reservations, they simply must put up with the growing piles of incomplete small, easy tasks. How to make everybody recognize that preparation is the necessary lead-in time to something important? How to persuade yourself?

SET-UP TIME AND BEING AVAILABLE

There are times when I'm sure I'm fitting in set-up time without quite realizing it myself. This is how it can look: a significant deadline is approaching and I know in my heart that I should have started the job long ago. Instead, I seem to focus on less productive things. I do nothing, effectively, and become preoccupied with pointless minor chores like washing up and mending and pottering in the garden and so on, even though I don't particularly want to do these things. I don't start attacking the real job until absolutely necessary, and usually a little later still. What a miracle that I meet the deadline after all – yet again! Or maybe it's not a miracle. I believe that by that late stage my mental workshop has already dealt with the task. Thinking and planning had been going on all the time my conscious self had been preoccupied with simple things. When the deadline loomed really large, very little was left for me to do.

Concentrated intellectual work demands a set-up time too, which might last only hours and days, or drag on into weeks and months. Once that time has been set aside, it must be properly used. You must be available within that timeframe only. To lock yourself within a certain task in this way is utterly contrary to the way we nowadays prioritize being available to all comers, be it by instant travel, mobile phone, email or whatever.

I tackled my relationship with the telephone some fifteen years ago, starting with my office phone. How was I to silence it? I could programme it to say that I was at a meeting or away on business or teaching or out to lunch or had left for the day. But the list did not let me say anything about the task I was primarily hired to do, like: 'I'm in my office/ at my computer/ in the laboratory – thinking' or 'with students or colleagues – talking'. I discussed my problem with a couple of switchboard operators and they told me that the message 'Bodil Jönsson is not available, she's thinking' would probably provoke an angry response. The caller would feel 'If she's only thinking, she might as well answer the phone.' This wasn't how I felt, though.

DARING TO BE A HERMIT

Working in seclusion is often important for good results. I have learnt this important lesson by now, and am quite capable of defending my need to live like a hermit occasionally. Trying to be truly present wherever I am is crucial not only to me, but to the people I work with. They should feel convinced that I'm *there*, with them. No telephone calls must be allowed to interrupt us. If it is your professional responsibility to think, it is indefensible to give in to either real or apparent demands and accept other, irrelevant measures of your worth. It is your fault if hackneyed and ill thought-out research and teaching come to dominate your output.

The set-up time for thought interacts with your environment. One example in my case is the effect of the Internet. I believe it has changed both my conscious and unconscious ways of working. Now I allow myself to pursue even more new lines of thought because of the ease and speed with which I can find information relevant to my ideas. This opportunity to search actively in many different contexts is fundamentally unlike the laconic statements in encyclopaedias and other reference books.

Encyclopaedias have a set-up time of their own, and are noticeably stamped with the imprint of their period. Their long set-up times mean that they rarely reflect contemporary breakthroughs or frames of thought. Information is mostly drawn from established sources – or in other words, tends to be out of date. This con-servatism is presumably necessary to prevent the inclusion of untried novelties that could turn the work of reference into a catalogue of ephemera. A ten-year-old girl once provided me with a striking example of this conservation of past-it information: her essay on Liberia described the country as 'populated by wild and semi-wild Negroes'. Baffled, I asked her where she had got this from and was told *Time's Encyclopaedia* – printed in 1938!

IS YOUTH THE SET-UP TIME FOR LIFE?

Now I'll reflect a little on the big subjects of childhood and youth. Both are often regarded as set-up times for real life. I find this a curious notion.

It is arguable that upbringing doesn't matter in little things like nice table manners. Sooner or later, people who need to know these things will pick them up. I believe the same to be true in other, more important, areas of life, though most people seem to disagree. Education tends to be seen as a benefit for later on. In our culture youth is a long-drawn-out affair that lasts much longer than our biological youth. It takes very different times to become regarded as adult in the social contexts of communities based on hunter-gathering, settled agriculture and industrial manufacturing. The successive extension of youth is justified as the time it takes to acquire all the knowledge presently demanded of adults.

What happens if we no longer get to grow into

adults? What will it mean to young people in the twenty-first century if their generation turns out to be the first to learn that adulthood is for ever out of their reach? This might well affect all later generations too. It would surely follow that childhood and youth, the most important parts of life, would finally be given the esteem due to them in their own right, rather than being seen as the set-up times for life.

My parents' generation still recognized a time of adulthood. They would have seen it as unacceptable foolishness for a thirty-five-year-old to dress in the clothes of a child and run about playing, or jogging as it's now called. Adults and children were distinct, and there was a consensus that children should be prepared for adulthood, and meanwhile be seen but preferably not heard.

How did the older generations come to have the upper hand in this way? Among other things, obviously the adults had acquired knowledge that was needed by the young. At the same time as the essential education in matters such as building, cooking and so on was going on, it must have been irresistible to instruct the young about the older generation's views on matters of religion and morals as well.

Now the situation has changed radically. Some jobs

require skills that the parents lack completely but the fifteen-year-old has picked up easily by playing computer games. It has become harder to argue that old folk know better and that youth is part of the set-up time for life. Today, no rule says that older people have more appropriate life experience than younger ones. Often the opposite is the case, because old patterns of thought limit our ability to experience new things. Even though what I wrote about the opening up of opportunities for older people is true enough (see Ch. 1), there are striking differences in the way older and younger people think, particularly in fundamental matters.

YOUNG 'THINKOLOGISTS' NEEDED

Above all, young people are needed as 'thinkologists' even more than as technologists. There has to be a distribution of responsibilities between generations, and I believe that the crucial role of the young lies in providing new patterns of thought, and also demonstrating the consequences of new thought. Respect for humanity and human activities needs to be based on new precepts and we, the adults, need the constructive help of those who – unlike us – have never been caught in the formal requirements that belong to the era of defining-your-worth-by-work. New experiences become distorted when they are understood in terms of old patterns of thought. The effect is like being in the House of Mirrors at a fairground.

An example of what I mean is our distorted current reaction to unemployment. How come we suddenly seem to regard work as a kind of desirable but strictly rationed good, attractive to all outsiders? Working is surely not a human aspiration since times immemorial? Rather the reverse, I'd have thought. I can't recall a single image of Paradise where people do anything but laze about blissfully.

The present form of unemployment, that is the shortage of jobs as formally defined, will not just continue but actually increase steadily, as far as I understand the situation. Of course there's no lack of real work. So it is important that a powerful group of clear-headed young people should be asked to help create new ways of thinking about both formal unemployment and the kind of work that is part of the reality of living. There is always so much to be done, and so much that it would be good to do.

In *The Emperor's New Clothes*, the classic tale for

children, a child tells the truth while the adults keep pretending. I believe that children are more likely to tell the truth than older people, but few adults are prepared to take children seriously. Teenagers take precedence, perhaps because of their height and noisiness. Another strength is their skills, notably in information technology areas: while the older generations still live either in towns or in the country, young people live in virtual villages.

MANY OLD OLD-BOYS AND NO ONE YOUNG

I keep thought-provoking images, and one of my favourites is a photograph taken at the global environment summit in Rio.* Every single person in the photo falls into the category of 'Old Boy'. It is age that concerns me here, not gender. Although it will not take many more years before half the population in the world is aged eighteen or under, the rulers of the world

---

* The United Nations took the lead in arranging this conference in 1992.

and its future are predominantly elderly men. It is absurd that this state of affairs should last for much longer.

I have another, related image. It shows the empty room where the Security Council of the United Nations meets, and it seems to me that the furniture itself sends a message of hope – of honest attempts to meet and talk, to mediate, compromise and negotiate new opportunities for mankind.

It will be exciting to be around when the young finally have power. Maybe that power will be different from that exerted in conventional governments of the past. Maybe they will create new power structures. Above all, it is essential that they do not see their youth as a period of indoctrination into established adulthood. The records demonstrate our failures, which include deteriorating natural and built environments, increasing discrepancies between wealth and poverty, endless warfare and the inability of our democracies to manage the market economy effectively. We need new guidelines for both thought and action, and can only hope that the young will not aspire to emulate today's adults.

SUMMARY

You need to let the concept of set-up time sink in, just as you have to become slowly aware of the idea of personal, lived-time. Once that's done, you should try to analyse ways in which your new view of set-up time will allow you to act differently from before. These are the perspectives from which I have viewed set-up time:

- The set-up times for hard and easy tasks are different, and affect the distribution of completion between different kinds of task.

- A period of apparent evasion and focusing on smaller, easier jobs can mask the actual set-up time of a harder task.

- Our environment, which includes the new IT-based means of information retrieval and communication such as the Internet, can have important effects on set-up times.

- Childhood and youth tend to be seen – mostly mis-guidedly – as the set-up time for adulthood.

4

Undivided and Sub-divided Time

*H*aving succeeded in my first, and so far only, attempt to stop time, I decided to change my delusions about life. I argued that 'I might as well believe that I've got plenty of time, rather than believing that I'm always short of time.' In other words, I decided to have plenty of time. This did not mean that I tried to get out of having any kind of relationship with time: that's probably not an option in our time-obsessed culture. You always have to decide whether you have plenty of time or effectively very little time – no compromise seems possible.

Does the exchange of one delusion for another really work – do I feel that I have more time to play with? Yes, it works, and the delight I feel at having enough time is neither more nor less real than was my horror at being short of time. In principle at least, the change of attitude has been easy, but the habit of feeling appalled at the swift passage of time is hard to break. Once that was done, it was as if time had become a huge gold-mine, an image much more pleasant than that of time as an ever-dwindling stream.

Try my new delusion and see if the notion of having plenty of time works for you. I found that, having persuaded myself, and gone on to tell other people, I really did seem to have time to spare. There are still occasions when I rush about like a maniac, but I'm always aware that it's easy to take a deep breath and restore orderly progress. Feeling you have more time is part of the orderliness. It doesn't mean that I don't do very much. I do far too much, but I keep the sense of time stretching into eternity by following a couple of simple rules.

One rule is always to protect my need for set-up time, both by planning for it and by defending it against encroachment. I respect set-up time and the agonies it entails. My second rule is based on the idea that life must not be sub-divided into too many little bits. At one point I was wondering why summer seemed to last so long when I was a child. The answer is that it stayed an undivided whole. No basketball camp in early June, followed by a trip to Majorca the first week of July, a mid-July visit to Granny, scout camp in late July and so on and on, leaving no intervals longer than a week or so. Everybody knows that a week doesn't last very long. It's a finite chunk of time. Adding up eight separate weeks holds no hope of eternal summer. The child's sense of endless expectation can only be re-created by leaving summer in one piece.

A continuous block of time leaves you space to live in. Timeless spaces, the exact opposite of your daily routine and a child's school timetable. In fact the school timetable prevents learning, by not respecting the pupils' need for set-up time and constantly interrupting their concentration on a particular subject (see 'Thinking Takes Time', Ch. 5).

TIME AND TERRITORIAL REFORM

Open your diary and look at how compartmentalized your time is. Consider your options for territorial reform, of the kind that means the pooling of scattered small plots to allow the cultivation of large, continuous fields. The main current challenge to us, individually as well as collectively, is to pool fragmented time and so bring about profound core benefits in our lives.

In both agricultural reform and reformed time, the idea is not to deprive any one land-owner or life activity, but to add value across the board by providing larger units. Most of us prefer undivided to sub-divided time. The difference between divided and undivided time is so great that both should not be measured by the same means – clock-time. Adding up measured chunks of time is rarely meaningful. For medium-distance travel, most train journeys are slower than flights, but a 3- to 5-hour period in a quiet train compartment is real, undivided time, whereas the 3-hour travelling time by plane represents the successive waste of three separate hours. The loss of time is due partly to the sub-divisions, but also to the many disturbances entailed in going to and from airports, queuing, listening to loudspeaker announcements, etc. All I want is peace and quiet to do what I like to do.

I might want to talk to somebody. Conversations should take place in quiet settings, though. To be allowed to focus on talking can be truly inspiring and a source of new knowledge, emotions and impulses. Most human beings have the capacity to develop and enrich each other. But to act as each other's mental midwives – to help each other deliver new thoughts and ideas – can be time-consuming. Interrupted deliveries are often inconclusive.

Consider the hours we spend trying to relax in front of the television. Some of us can't stand the interruptions of advertisements, and feel happier with public service channels, even though the constant bombardment of announcements and trailers can be just as irritating as the advertisements. Is real relaxation with TV an impossibility?

A POSSIBLE ROUTE TO
INDIVIDUAL DEVELOPMENT

Your route to a less fragmented time could look like this:

- Become aware of the difference between undivided and sub-divided time; unless you understand this you'll remain unable to change the balance between them.

- Stop regarding undivided and sub-divided time as the same kind of time, measurable in the same way.

- Try to reform the way you manage time by pooling fragments into bigger blocks of time.

- Try to achieve the same time-use reforms within the social groups you're part of.

Work out how much undivided and sub-divided time there is within the different timetables that govern your life, be they associated with leisure, work, school and other children's activities, holidays, entertainment, TV watching and so on. Make deliberate choices, aware of what each activity means for the sub-division of your time.

ADVANTAGES OF SUB-DIVIDED TIME

Now I'm going to make a U-turn and argue in support of those who feel that sub-dividing their time is necessary to their sense of being in control of their lives. Many people *need* to dissect their time in order to grasp how their world works. Children want to know things like 'How many soons make a while?' and enjoy explanations like 'It'll last for half a comic.' Some adults also find precise time indications invaluable in order to be able to understand change at all. In such cases, time-lessness is far from a desirable dream.

The case of the structure of Henry's mind is interesting. Henry was a mildly learning-disabled, 50-year-old man with a vocabulary of perhaps a hundred words and phrases. He was introduced to a personalized electronic 'assistant' through which images of Henry's life were stored and could be revisited. He liked the images, but initially found them hard to relate to unless printed. Once several of the prints were displayed on the wall he was seen wandering up and down attentively observing them, mumbling to himself all the time, somehow speaking to or about the pictures. Standing next to him one day, I realized that through these images of himself, time had become part of his life in a new way.

I believe that in his previous state of mind, Henry had been short not only of words but also of remembered scenes. His mental rooms were bare, and it was hard to start furnishing them without any precedents. As a result, Henry's life had been one long succession of 'nows'. To wish, to compare, to expect – all these and many other mental processes were practically impossible. Thought is closely intertwined with retained experience.

Because learning depends on being able to play with

variations on a central theme – a memory sequence – learning too had become very problematic for Henry. In spite of the widespread belief that repetition is the key to learning, it is in fact only a partial truth. Repetition is an aid to focusing. By repeating we get a stable platform for the constantly changing perspectives that are essential to real learning. In the past, Henry had found most repetition without effect, and so variations were meaningless. He had no active memories from his life. Through the images, Henry's past became part of his present, and an inner life began to flourish. These scenes had obviously always been stored in his mind, but he had been unable actively to recall them. Now they were there, tangible evidence of yesterday to be compared with today. It didn't take Henry long to begin wondering and wishing about tomorrow.

Now an explosive learning process got under way. The first sign was increased alertness, soon followed by a huge expansion of Henry's spoken language, with regard to both passive and active vocabulary. Initially, this was most obvious in the context of the images, but it seemed that consciousness of time was linked to his mental development.

We all depend on being able to anchor the present in both 'yesterday' and 'tomorrow'. Memories become meaningful in proportion to our ability to order them in relation to a before and an after. Without such ordering, life would become chaotic. We need to know that 'it happened before we moved', that 'it's what he did after shaving off his beard', etc.

One group who I think would find sub-dividing time very useful is people with early signs of dementia. In old age we often almost entirely stop taking snapshots, yet photographs are such a help to the memory. Baby's first year is usually well documented, but then the pictures become fewer and fewer, until for fifty- to seventy-year-olds there are hardly any records at all of recent events. Digital photography and video mean that it should be relatively easy to create a TV channel based on your own life. On this 'Sweet Memories' channel, you could watch both the recent and the more distant past. Seeing yesterday's events and being reminded of the plans for tomorrow might help reduce, or at least delay, the crumbling of the memory-structure at the core of the self.

OLD-TIMERS:TV AND THOUGHT

Our observations on how other people use their time can arouse absurdly mixed feelings. The way elderly people live is a case in point – or rather, what old folk get up to in their presumed loneliness. The author of a recent thesis on 'Being old', discusses our misinterpretations of what old age can mean.*

For instance, to feel better about ourselves we worry about not being with older relatives and friends as much as we should, or not visiting them as often as we can. It doesn't occur to us that many old people are not that keen on being visited and that what they like is to contemplate life in peace and quiet. A preference for sitting and thinking has come to seem unacceptable, and TV provides a handy alibi. When there is no one to argue with, no one to say 'Come on, you're not listening', quiet meditation, remembering and occa-

---

* The author of the thesis is Britt Östlund, Department of Technology and Social Change, University of Linköping, Sweden; the department runs a programme called Tema T, aimed at 'understanding humans and society via their interaction with technology and analysing technology in relation to social change.'

sional watching can easily be interleaved. For many people, the TV set is a less disturbing as well as a more tolerant and amusing partner than relatives and friends. Besides, it is a useful means of getting chunks of undivided time.

IN PRAISE OF UNDIVIDED TIME (AGAIN)

There are people who find it helpful to sub-divide time: to dissect it into its parts as an aid to seeing a coherent structure. The rest of us, who could be called people-in-the-middle-of-the-rush, must not assume that our wish for undivided time is universally felt. Life as a smooth, unending stream is a nightmarish vision for some.

But sub-dividing is not a totally good thing, even for those who have difficulties with remembering. Many in that group also find it problematic to switch between activities. Changing from one activity to another can need so much mental preparation that the activity cannot be enjoyed. Only the set-up time is left, and the result is just confusion.

Maybe that's how you feel right now. You have followed my reasoning, and just as you begin to feel you've caught my drift, I come at you from another

angle. That's life, you know: multi-faceted. Not complicated, but complex.

THE COMPLEX VERSUS THE COMPLICATED

I very much enjoy finding difficult – complicated –
issues, sorting them out and, from time to time, presenting them to others in a more easily comprehensible way.
Simplification can go so far and no further, of course.
Some relationships are so complicated that attempts to
simplify them turn into metaphors or just fairy-tales.
But some complicated matters can be unravelled like a
tangled skein of yarn, if there is one free end to follow.
It's nice when you end up with a neat ball, whether
you're dealing with skeins of thought or yarn. It is useful
to make complicated matters simpler, if at all possible.

Time, and our attitudes to time, are usually not complicated but complex. Trying to simplify complex
matters gets you nowhere. Eliminating complexity is an
attack on the core structure that only leads to damage
to the whole. If a complicated matter can be likened to
a tangled skein, a complex one is like a tapestry: pulling
out threads of warp or woof removes something
essential from the patterns and shapes.

This is true of time. Without respect for its complexity you will never understand anything about it. Don't try to simplify. Complex matters must be studied and observed over time. In all the examples I give in this book, the essential thread is that it takes time to think about time.

5

TTT – Thinking Takes Time

*P*iet Hein, a Danish writer and a 'philosopher of little things', coined the phrase 'Things Take Time'. My own variant is: 'Thinking Takes Time'. It takes time to create new thoughts, develop them (I call it 'Thought Care') and then get rid of them when they're out of date. Even little thoughts take a finite amount of time. Dismantling systems of thought and then rebuilding or replacing them demands thinking of the most time-consuming kind. Thought leads to social systems, and nothing dictates the future as ruthlessly as an established thought infrastructure. The most effective way of changing the future is to create a new system of thought.

One system we must replace as soon as possible is

the one we have inherited from industrialization. Its central notion was that work was crucial, and what you did when you were working mattered too. Being employed was synonymous with being needed. Leisure time was when you did things like entertain yourself or look after your private life, including children and older relatives, and chores such as cooking, cleaning and laundry.

The post-industrial society must get rid of this work-based outlook. The hard bit is managing to break with the prevailing trend. All such disruptive changes have in common at least one effect: they upset people. This is what happened when we stopped putting a social premium on the drainage and cultivation of wetlands and instead began celebrating restorations of marshes and water-meadows. Many of those who had spent a lifetime digging ditches and struggling with damp soil reacted violently. Their mixture of anger and guilt was perfectly understandable. The storm of anxiety-driven rage that will greet us, if and when we decide to break with deep-rooted behaviours such a personal car transport, are just as predictable. But it'll be a breeze compared with the outrage that will confront us when we launch a new definition of what constitutes a useful citizen.

THE NEED TO BE NEEDED

It is generally accepted that people can be needed for different reasons. We like to feel needed even when we are out to enjoy ourselves. Yet in most cultures and most periods of history, there is evidence of a human desire to decorate our bodies and go dancing when not required to work – dance not just for our own sakes, but for the sake of the feast, the gods, the powers that be. Without always being literally true ('dancing' can take many forms), this idea is based on people's need to express themselves, often joyfully, in the company of their fellow men. Of course, that is impossible without shared systems of thought.

One effect of information technology is to make us ever more independent of others – all of us individuals living in different worlds. How to foster a culture of sharing? Nowadays it is possible to share a common cultural heritage without a shared vision of the future of that culture. I think so, at least. I also believe that nowadays we often experience a collective sense of loss, caused by that lack of a common goal, and maybe even of a common faith.

It is hard to see how such a vision could be constructed. Sometimes I feel like the three-year-old who had been talking on the phone and got round to saying 'Bye, bye'. The kindly adult at the other end asked 'Don't you want to talk any more?' and the child replied 'After bye, bye there's no more talk in the phone.' The logic works at the level of a three-year-old mind, but is certainly not acceptable for adults. Or can it be that we feel that we have run out of 'talk in the phone', that our common future is not an appropriate subject because we're all individuals now?

ALL THESE FADS

Human faddishness is remarkable. Fads can get established very quickly, but it takes an age to understand them. Our love of cars adds up to something different in kind from the composite total of all cars and the road network. The combination amounts to a way of living and a way of organizing economy, trade and social patterns. Gradually, car culture has come to dominate thought patterns too.

I belong to the generation born in the 1940s, and when I was growing up 'car-society' became established. My father bought his first car in 1947. It was a great event, not only for him but also for the whole village.

He wanted to take the whole family out on a trial run, but I wanted to play with my bicycle instead. I had modified it by sticking a piece of paper to the rim of a wheel to make it rattle as I cycled – sensational! I never heard the end of how I refused to go in the car because I enjoyed my noisy bike more.

I have always been a contrary person, and still am. It doesn't mean that I haven't joined the car-owning society. At least, I own a car, though I prefer not to use it except when necessary, and I still get great pleasure from using my bicycle. Above all I enjoy the cycling tempo, which I feel is just right for taking in impressions of the world around me. But the car controls my world to an awesome extent. Cars are the essence of so much that happens around me, and mean so much in my day-to-day existence.

The next big fad is information technology (IT), and it is advancing on a broad front. IT is all over our news-papers and magazines, even though it is such a recent arrival. So is the rule that 'Thinking Takes Time' not valid for IT? Both the ideas and the practical applications have been flooding society so fast that it is tempting to conclude that Information Takes No Time (ITNT). Thought processes can't quite keep up, and new thought systems to incorporate these new modes of operation are still slower in coming. ITNT signifies a kind of mental supermarket, while old TTT stands for The Temple of Thought.

In time, the enthusiasm for IT will end up as something quite different from the sum of individual modems and computers, e-addresses, websites and the rest, just as the whole car-owning society turned out to be something more than the sum of the cars and the roads. By the same token, IT society will not be 'more of the same' but 'different'.

Faster is not just 'faster', it is also 'different'. The changes related to IT will just happen more swiftly than the changes accompanying car ownership. Speculating that the relationship will be 1:10, and setting the time taken for the car to change our lives at fifty years, it shouldn't take IT technology more than five years to restructure our society for good.

TELE-PERSPECTIVES AND OUR ENVIRONMENT

Many of our environmental problems are by-products of our current preoccupation with being short of time. It has become part of contemporary culture that we lack the one asset we all have a share in, as I argued earlier. Mankind's future status as a natural resource for good or for evil will depend on how our relationship with time develops.

One of our cherished beliefs is that the concept of distance has become obsolete. We are convinced we can handle global interactions by using telesystems. In order to compensate for our deficiencies, we have actually become very good at 'tele'. Human hearing has a limited range, so telephonics (remote-sound) was a useful invention. So was television (remote-sight), a technical device making up for our short-range eyesight. Apart from telephones, televisions and other tele-gadgets, we can now also indulge in tele-consumption, and as a result, produce mountains of tele-rubbish, i.e. solid waste of all kinds, to add to our discharges into air, land and sea.

There are other aspects of tele-culture that make me anxious. I'm not sure that tele-knowledge is all that enabling. Information, yes – but knowledge is far from being the same thing as information. Interactions between human beings cannot easily be mediated at a distance. Is it possible to become 'tele-wise'? To negotiate 'tele-peace'? To shoulder 'tele-responsibility'?

FAST TELE-REVOLUTIONS AND SLOW LOCAL CONSERVATISM

Local systems of thought can take much longer to change than global ones. Compare, for instance, the rapid revolution of international economic systems with the stubborn refusal to change old measurement units in the UK, USA and Canada. In spite of grand assurances, there is still a lot of resistance in practice to adopting the metric system.* Old habits go deep; inches, yards, etc. practically seem to be part of the genetic information in these countries. Individuals, however internationally minded they claim to be, are often deeply conservative about their home and hearth.

SLOW MEALS AND FAST FOOD

Food is central to our lives, and it is surely significant that the slow meal is increasingly being replaced by fast food. In the process, we have been deprived of so much:

---

* Also known as the SI-system, e.g. metres and kilograms for measuring respectively distance and mass and so on.

things like the smell of freshly-baked bread, interesting flavours and the charms of a well-used kitchen. Above all, the rhythm of our daily lives has changed. So the slow meal takes too long to prepare? Not enough time to wait, no snacks to still one's hunger quickly? It might have felt downright boring, but waiting was a good antidote to feeling harassed by lack of time.

GLOBALIZATION AND SPEED

But let's leave the kitchen and go out into the great big world outside. One fine summer evening recently, a professor of ecology was heard to say: 'I think we've been thoroughly cheated!' His comment started a conversation, not about simple solutions to problems or how much better things were done in the past, but on the question: 'Where is the evidence of progress and how do we find out about it?' The professor's first contribution was to talk about the situation, as he remembered it, in the small industrial town where he grew up in the 1950s. It had a school and health centre, and people led lives which were in many ways very satisfactory, but cost much less than corresponding modern lives do.

So had things really improved? Where does all the money go? Could it be that the monetary system is not single but dual: one fast and one slow method of exchange? And, because 'Thinking Takes Time', has this division passed us by?

In the not too distant past, money was a means of facilitating the exchange of services and goods. You went to work, got paid for it, and used the cash to buy food and pay the rent. But quite recently – maybe just a decade ago – that secure system of exchange was undermined by another, belonging to the new global networks. These networks made possible the exchange of virtual, electronic currencies without any direct relationship to goods or services. In this brave new world, actual financial transactions are no longer essential, expectations are enough. It probably won't take long before the old trading certainties have disappeared completely.

Arguably, money still has a geo-political base. A crisis may be located in a specific country, and the cause may lie in declining confidence in the financial strength of that country, or indeed continent. However, such thought-patterns belong to a past when currencies had real values. Quantifiable relationships of this kind disappear into thin air once the wheels of commerce start to turn freely, without the braking effect imparted by friction with the real world. In other words, the friction that makes possible important velocity changes such as leaps from a standing start, acceleration, steering and stopping.

Inertia was an inherent part of the old exchange system of money, goods and services. Producing goods and employing (or even sacking) workers take time. There are no time delays in the new system. Money is exchanged for money, and the transactions take mere fractions of a second.

It seems to me that in addition to the Security Council, the United Nations needs a council designed to deal with financial security. More than a European Union we need a Global Union – a GU. Our lifestyle has made it essential to keep global systems under control, and governments must allow for management at the levels where the effects of our capacity to change are actually visible. But it is also clear that we find the option of a global ruling elite impossible. Many of us are reluctant to accept being ruled from a European centre. It has not escaped me either, that there is sporadic conflict between national and European leaders.

These changes in the economic systems of the world are genuinely new. The same is true of the changes in environmental development. Current events have no parallels in geo-political history. For the first time, some changes in the economic and ecological balance are truly global. True in real life at least, though maybe not in our minds. Not yet.

DEEPER DIVIDES

The world is full of injustice. The divides are unbelievably deep. Worse still, and here time becomes significant, they are growing deeper. Their growth-rate is accelerating beyond our ability to follow and understand the process. This acceleration prevents a sense of solidarity between rich and poor, between communities, cultures and nations. Organizations as well as individuals have not enough time to get their bearings in this new world. When something bad happens, the source of the negativity is usually displaced to a diffuse 'somewhere else'. Always elsewhere, never here: the result is that suspicion and hostility towards the unknown are increasing too. And so here we are, ever more alienated in a world full of our tele-equipment. Goodness, Thinking Takes Such A Long Time!

The divide between generations is also growing.

Suppose that it is correct to say that the rate of innovation has increased by many orders of magnitude over a comparatively small number of years: the consequence would be that for people born between 1995 and 2000, technical change would be the equivalent of that occurring over several generations in the past. What does this mean for different professions? Just consider the teaching profession. Is it not the case that teachers age much faster now, from the perspective of the pupils they teach?

TTT – MAYBE MORE USEFUL THAN NOT?

When I summarize the headings of the sections in this chapter, the list looks quite melancholy:

- The Need to be Needed

- All These Fads

- Tele-perspectives and Our Environment

- Fast Tele-revolutions and Slow Local Conservatism

- Slow Meals and Fast Food

- Globalization and Speed

- Deeper Divides

TTT – THINKING TAKES TIME

It is a great shame that we are so slow to grasp what is happening, and therefore unable to stop mistakes being committed in the name of development. But still, maybe we should be pleased that our thinking can't keep up with the pace of change. It could be that this imposed delay is our best collective life insurance.

6

BEING PRESENT IN TIME AND PLACE

Being truly present in time and place – belonging in the here and now – can be an overwhelming experience. I remember coming across a simple engraving by the artist Helen Plato, showing a couple of standing columns. At first, I thought it showed a place. I was intrigued and looked more closely: its title was *Present Time*. Now it hangs just inside my front door. When I see it, my first thought is always: 'Here I am. Now!'

It is actually quite rare for one's awareness of place and time to connect in this way. Returning to places one knew well in the past can be disorienting: 'Odd, this rowan tree has lasted all of fifty years and now it's dying, suffocated by lichen – and it's happening so quickly.' On the other hand, it's possible to feel close to someone even when the two of you are far apart. Sometimes – but not always – you feel present within yourself, irrespective of place. Presence in time and place do not always happen at the same time, which raises issues that are worth thinking about.

Mobile phones disrupt the connection between present time and present place. I remember a TV advertising ditty that went: '*Here* is with you wherever you are...' True enough, I suppose. It would also have been true to sing '*Now* is with you wherever you are.' Anyway, one always hopes to be aware of one's own Now.

My need to feel as undisturbed as possible in my own Now means that I prefer post and e-mail to the telephone. Nowadays it has become easy to be relatively independent of time. It was different once, in the days when one person's actions tended to have immediate and decisive effects on what others were doing because people were sharing each other's present.

Space, too, used to be a fixed commodity. A few generations ago, most people could have at best only minor effects on what was going on in places a few kilometres away. A local failure to cultivate the soil properly would be the individual's own problem. By the same token, changes in distant places did not impinge much.

Assume that a typical person today travels an average of 50 kilometres per day. Distant places have come to seem close as the means and speed of transport have multiplied. Maybe it will soon become possible to feel aware of a global present, in terms of both time and place. But I don't believe that people are yet able to process perceptions on a global scale.

The arctic tern might be, though. Terns migrate across the Earth from the Arctic to the Antarctic, not once but several times during their lives. Every tern eventually covers a total distance similar to that between Earth and the Moon. Presumably a global sense of place is essential for this kind of thing. But human beings are not designed for travel between the poles. Some people still feel disorientated after being taken just ten or twenty kilometres away from their home territory. We need technology to reach far-flung places, and it actually suits us better to stay put within a geographically limited area. Even when we do use machinery to facilitate our travels, there is a limit to how far we want to go.

A PLACE OF ORIGIN

I'm talking to a three-year-old grandchild: 'I'm your Mummy's Mummy, don't you see? When she was little she lived with me.'

'Ummm. But where *was* I?'

'Oh, you weren't there yet. It was *her* dolls' house then.'

'Yes, but *where* was I?'

'Listen, when she was as little as Beatrice is now, she couldn't be your Mummy. You weren't around at the time!'

'But… where *was* I?'

I like that child's stubbornness. He had no problem with not having been born once – but where was he at the time? His ideas of time and place had proper shape and were practical and logical. His question had not occurred to me before. I had, however, thought about it the other way round, as it were: to what extent do unborn children affect our actions in the present? I don't mean foetuses and babies just about to be born, nor even our great-great-great-grandchildren. It is the totally unknown, unborn children of the very distant future that preoccupy me at times. Anticipating their existence affects our present-day actions. I had never managed to think the same thing backwards. To do that, I needed the help of a small child.

Perhaps it is precisely the feeling that 'I must have been somewhere before I was born' that has led to the increasingly widespread interest in genealogy. Knowing one's family origins is a way of somehow being present in the past.

THE CONCEPT 'HERE'

In my view, people are present in both the 'here' and the 'now'. A human being is very much a localized phenomenon. Everything he or she does can be defined as happening in a certain place at a particular time. It follows that human networks must exist in the dimensions of both time and space. 'Here-ness' has its limitations. I speculated about the possible contexts of the word 'here' one summer's day when I was picking blackcurrants. It was obviously right to say 'here, among the currant bushes' and also 'here, on this farm'. I can say 'here, in Lund' to describe being in my hometown, and also use 'here' for the region and even country where I live. But I draw a line at 'here, in Europe'.

This doesn't have much, if anything, to do with the fact that my country only joined the European Union recently. Of course this means that as citizens in a

European member-state, people are now free to travel without passports, exchange goods and services without hindrance and agree more and more often about customs and languages. Soon we will all use the same currency and maybe share policies on matters such as foreign affairs and defence. All this is still not enough to make me feel able to say that I am 'here, in Europe'. I believe this is because the mesh of my European network is too large: my contacts are too few and my visits too sporadic. 'Network' is a concept I find crucial to understanding the sense of being present in the here and now.

BEING PRESENT WITHIN A NETWORK

Everybody has a network of his or her own. A very young child enters into a family relationship with a small group of people who form the child's first, personal network. These are the people who will influence its new personality. It will become imprinted with the way these people live, behave towards each other, approach and solve problems.

After the early 'personal childhood' comes the 'cultural childhood', during which the main influences come from schools (from nursery classes onwards) and leisure activities. Now the network is growing all the time, and so the space for the individual's personality is shrinking, at least in one sense. The most important task for the family adults is to provide stable links to the personal childhood. Without such links everything new might come to seem a pointless add-on. The worst thing that can happen is for a young person to feel excluded and anonymous in a world where other people are interchangeable: it is even possible to feel that one's own self is mutable. Of course Miss in primary school will change from time to time, but generally the people close to a child should not come and go too often. At least one adult should be there all the time for the child, who demands a presence of familiar 'others' in the here and now.

Without such firm anchoring points the child easily slips away and starts drifting, and later may become antisocial or even violent. When the notion that people don't matter – are interchangeable – has become established, it's no big deal if someone gets killed. The individual has become faceless. On the other hand, if the self and other people's selves are seen as unique, it is clearly not valid to regard people as interchangeable. Instead, every person you meet becomes someone with a distinct identity, existing in a personal space and time, and with the potential to become part of the networks linking people with each other.

BEING PRESENT IN TIME AND PLACE

THE SELF AND THE IMPRINT OF OTHERS

The environment always leaves impressions on the minds of human beings, but it is during childhood that this process, called imprinting, has its most profound effects. Because I grew up in the country, the mesh of my own childhood networks was large. The people involved were few, but remained important to me for a long time. My image of self was very much influenced by how these people saw me. Their reactions to me in words, body language and actions – the whole human interplay between them and myself – were crucial factors in my development. Most children will have formative experiences of this kind. It is this human interplay that shapes our culture, our sense of 'being here'. Togetherness lies at the heart of all culture. The reverse is also true: rejection and abandonment is part of cultural breakdown.

BEING PRESENT IN A CROWD AND IN SOLITUDE

'Being alone, now – you know, Bodil, loneliness is the worst enemy of human beings,' my Uncle Erik once said to me. He was standing in the yard of his smallholding and autumn was in the air. I thought I could hear in his voice the approach of winter, snow and isolation. And I think I understand exactly how he felt, even though I'm a semi-hermit and instinctively search out solitude. But someone who does not have a network within reach is likely to feel his or her self-image crumbling.

Of course people are more scattered in the country than in towns and cities. Globally, the trend is towards increasing population densities in cities. In my fairly typical northern European country, 83 per cent of the population lives in towns or cities, where 9 of every 10 children grow up. It is not easy to credit, given the way we all rave about 'returning to nature' during the summer. At the millennium, more than half the world's population lived in cities, whereas at the turn of the previous century it was only 12 per cent. In 2000 only Tokyo exceeded 20 million inhabitants, but by 2015 the world will have at least seven cities as large as that.

In just 100 years the proportion of city-dwellers has gone from 12 to 50 per cent! What is dictating this degree of urbanization? Perhaps a combination of the three strongest driving forces known to man: laziness, greed and selfishness? Are these what we're looking to gratify in the diversity of big cities? Or is it the dynamism, the beating pulse of city life? We want it all – and we want it now! Maybe it's the exciting rhythm rather than the density of the human networks that attracts us?

I don't know the answers to these questions. What I do know is that the strength of driving forces can be independent of their effects. Maybe the less dynamic but more permanent networks of childhood led to more durable results, including an easily identifiable self-image? No one is in a position to make oracular pronouncements about what is right for different people. There seems no good reason to apply the brake to the accelerating shift of population from the countryside to built-up areas.

But in just a few generations we have fundamentally changed our children's living conditions: how they grow up and how they form their sense of self and time.

BEING PRESENT WITHIN OVERLAPPING NETWORKS

The smaller the country you live in, the greater the likelihood of overlap between individual networks. It is just such overlaps that confirm who we are and that we belong to the same community and the same culture.

Whole societies would collapse like a house of cards

if these overlapping networks did not exist. One recent example is what happened to the social structure in the Eastern European states. Human networks had shrunk steadily because people never knew whom to trust and who might be an informer. Restrictions on travel also contributed. With time, the overlaps had become so small that the innate stability of the national unit weakened, and could not sustain the tottering and much hated regimes. A wind of unrest was blowing, and winds knock over houses of cards.

The Internet is of course providing overlapping networks of unprecedented dimensions. While the overall net is gigantic, it cannot become too large as long as each user creates his or her net according to personal preferences. The outreach of each person need only be as large as seems right to them. Your own Internet horizon is very much a movable feast, and you need only invite the companions you want.

BEING PRESENT IN YOUR OWN MIND

There are no researchers, however keen, who have answers to all the new questions that turn up. The realization that most people are perfectly able to think for themselves is reassuring. The best strategy is surely to encourage everyone to try to solve their own problems and look for answers to their own questions. The worst of all possible strategies would be to fashion a society whose members lacked freedom of thought and action – or, in other words, were just hanging about waiting for a strong leader to come along and sort things out. There will always be someone who insists that he or she has all the answers. It is when everyone believes in the same guru that things can become really dangerous.

The opposite is a crowded scene with lots of people in different activities and belonging to different networks. Usually there is a stable point in a dynamic system, a restful place at the centre of the movement. Even though we feel that change is taking place too rapidly now, I'm sure that it is still slow by comparison with what the future will bring. That's why we must try to find rest by being present within our own minds.

People are already saying that they feel exhausted by the swift pace of change. As it accelerates, we will become increasingly unable to alternate between running to keep up and calling halts, telling ourselves 'I've had enough change now to last for a good while; so far and no further for the next couple of years at least.'

Huge leaps followed by moments of almost obsessive stillness are in fact nothing but jerky movement. If the rate of jerkiness increases, it will sooner or later become continuous movement instead. Then we will be forced to try to find a way to rest while moving rather than by being still.

DOES 'SPACELESSNESS' EXIST?

The existential questions about life and death are timeless in the true sense of the word, unlike so-called 'timeless fashions'. Fashions are never really timeless. Some might last for a decade or two, but hardly ever for a century.

What about something being 'spaceless'? Is such a quality possible? Well, look at the miraculous way we've practically eliminated distance. It has taken us just fifty years or thereabouts to reach a situation where we count it in hours instead of miles. The distance between two places is less important than how quickly you can get from A to B. 'Here' and 'there' are about to lose their meaning as geographical concepts. Yet human beings remain localized to a very high degree. The question is, can we use concepts other than distances, surfaces and volumes, i.e. human-related dimensions that we can grasp easily?

BEING PRESENT IN TIME AND PLACE

Our thoughts can be spaceless. We stop using distance as a measure when we think. We think we can understand, for instance, the natural environment without referring to a particular space. Usually, we prefer not to speak about our spaceless, placeless modes of thinking, because someone would be bound to start a sermon about how bad it is not to take regional and local characteristics into account. Of course, this happens all too often, as cases of big construction companies imposing 'solutions' without consulting local people show.

HOW DO YOU CREATE A PRESENCE IN TIME AND PLACE?

In his autobiography Bertrand Russell wrote:* 'Three passions, simple but overwhelmingly strong, have governed my life:

• Longing for love;

---

* Bertrand Russell (1872–1970) won the Nobel Prize for literature for his *History of Western Philosophy* and was the co-author of *Principia Mathematica*. The quote is from his autobiography (1967): *The Prologue – What I Have Lived For*.

- Search for knowledge;

- Unbearable pity for the suffering of mankind.'

I believe that you will find yourself more fully present in the here and now, the more completely you manage to share these passions of Russell's. If they set you goals that prove elusive in reality, then at least for people with imagination there is always the option of entering into an inner space. Those around you rarely appreciate this. From early childhood onwards you're told: 'Snap out of your dreaming!' Often the ability to withdraw into yourself is a precious gift, and far from a behaviour that should be suppressed. A child who has been allowed to dream will later in life find it easier to accomplish the three important tasks of looking for knowledge, longing for love and empathizing with the suffering. These all involve acting in the outside world, and the child must be allowed to wait until he or she feels able to begin. In the meanwhile, dreaming sustains the mind.

Apart from mass meetings, rock concerts and sports arenas, is it possible to find a kind of collective sense of being present in the 'here and now'? I once heard a talk by an Indonesian man called Gede Raka about the essential features of a creative environment, and suddenly I recognized his prescription:

- Learning-friendly

- Oriented towards doing good

- Friend-friendly.*

Even though the items in this list had been formulated in an utterly different context, they are plainly very similar to Russell's insights.

My contribution is a single sentence: *Being present in space and time creates a framework for personal creativity.*

---

* Prof. Gede Raka is at the Center for Research on Technology at Bandung Institute of Technology, Indonesia. I have not found the source of the three listed criteria, but some of his thoughts on creativity are set out in 'Stimulating grassroot creativity for quality of life and quality of Environment', *Water Science & Technology* Vol 43 No 4 pp 167–173, 2001.

7

AWARENESS OF TIME
AND THE PACE OF CHANGE

So far I have been insisting that human beings are not very good at measuring clock-time. I have kept referring to the other kind of time – our lived, personal time. Your experience of lived-time depends less on absolute units of measurement and more on the quality of the time, i.e. whether it is undisturbed, how much is needed for set-up time, and so on. But it is a fact that people have built-in timers, as have plants, fruit flies and bacteria. Before we acquired the skill of making clocks we had to rely entirely on our internal clock-mechanisms. Now our internal timers co-exist with external clocks, and one consequence has been that we have discovered just how adaptable our physiological timing system is. Our adaptability is one of our greatest assets but also one of our greatest weaknesses.

If we construct a technical device that operates on a time-base different from our own ancient one, we quickly adjust to its demands even though we are not suited to the new conditions. Jet planes create fast changes of time zone, and we adapt to this. After a few days of jetlag the body settles down to the new time regime. We assume that driving a car at seventy miles per hour is a perfectly reasonable thing to do, and see the landscape outside as static: i.e. we have completely discounted 'speedlag'. We also seem able, after a brief 'pacelag', to adjust our sense of time to the pace of change in the world around us. That last observation is the main subject of this chapter.

It's often specific technical inventions which form the basic conditions for more general change. The coincidence of events such as the fall of apartheid in South Africa and of the wall between East and West Germany is far from a coincidence. The media that helped bring about these changes continue to monitor the activities of those who want to resurrect the bad old days. CNN, the US news broadcasting company, was in place to report the attempted coup aimed at ending *glasnost* in the former Soviet Union. What would have happened if the TV company had not been there, as it could only be thanks to the freedoms already gained? By the same token it was predictable that the world economy would create global markets once a global electronic network became available to all, including currency dealers and their colleagues. In the educational world, it is unsurprising to find students in mental search mode, rather than ready to receive knowledge. There are enough technical devices available to assist their search now.

PACELAG

I can recall the first landing on the Moon in 1969, but experience the time as almost unbelievably long ago. I had in fact already been around for twenty-seven years by then, which is almost half my life to date. If I try to think ahead, I'm aware that no one alive now can even begin to guess what will happen in the future. I know only one thing for certain: the changes in the coming twenty-seven years will be even greater than those that have taken place during the previous lot.

How will our civilization deal with this? Everything

is moving faster and faster. People are beginning to react. We don't like it. Our bodies protest. I believe that there is an antidote, and I've already suggested what it is: to find some kind of rest within the constant movement.

Perhaps I should start by having a look at the classical Greek ideas, though they were apparently not interested in movement per se. The contemporary view was that bodies in motion, subject to changing circumstances, could not be investigated precisely because at any one moment, by definition, their properties are different from the moment before and after. In spite of the gulf between our systems of thought, there are still classical ideas that have inspired me to try to apply them in the context of change.

Plato insisted that an object on Earth – a horse, say – is less important than the general idea of 'horse'. It's the 'horse-ness' of the horse that matters and endures for ever, while any present horse is but an incomplete shadow of the idea. Imagine looking at technology in the same way. Then we would look for the ideal machine, the mental image of which is represented in faltering replicas we have created in all our gadgets and gizmos.

THE IDEA OF THE CLOCK – 'CLOCK-NESS'

A clock, now – what's it for? Once upon a time clocks were made to reflect perceptions of cosmos. This was true of watches as well as of cathedral clocks and sundials. Clock-time was a calm, reassuring measure of the sun's path across the sky, seen from a human point of view. These early clockworks were replaced by ever more sophisticated devices, ending with our current best, the digital clock. One of its characteristics is that it tells us nothing whatsoever about the cosmos.

Digital clocks are symptomatic of a drive for precision that is relevant in both micro- and macrocosms, but can tell us nothing about our cosmos. Is the digital clock a symptom of a break in the continuity of ideas, much more profound that it might seem? Maybe now the idea of the clock, the clock-ness of clocks, which lies in its ability to show us the passage of natural time, has been replaced by something that only indicates an artificial kind of time?

INDUSTRIALIZATION AND THE CLOCK

Why do we chase time the way we do? One crucial reason is that we have accepted the thought-patterns of the industrial era. It was industrialization that placed us under the domination of artificial time. As I have discussed already, its dominance and its impending collapse directly affect our attitudes to work and the value of work.

A century ago, people fought against having to work themselves literally to death. Then, the notion of trying to create more jobs would have seemed absurd. The goal was freedom from work. 'Unemployment' would have been seen a cause for jubilation and raised questions like, 'You really managed it by 2000? No one kills himself working and you can draw money without working?'

But then as now, the human wish to feel needed was fundamental. Sometimes I think we should create a Ministry for Unemployment, headed by a full Cabinet Minister. His or her job would not be to eliminate unemployment, of course. Instead the Minister would be responsible for seeing that unemployed people are needed and actually feel valued. There is so much to do, but sadly no rule insisting it must be done according to new norms.

CONDITIONING

We are left with a relic of the industrial era: the dominance of the clock. The way we learn about clock-time is reminiscent of conditioning in many ways. We react instinctively to the clock and its rule over our time. We are conditioned to start work when the clock-hands tell us the time has come.

The peak of industrialization has passed, and it is probably true that soon manufacturing will employ as few people as farming does today. The related patterns of thought are fast going out of date. Some still hang on, including our victimization by the clock. There are other parts of our lives that are also still in thrall to industrial discipline. Just think of the vocabulary of work-rhetoric: efficiency, rationalization, work versus leisure time, and so on. These cannot be automatically applied to working with people, such as in care and education, and are also mostly useless in the context of inner work, such as learning and research.

It is still more pointless to try to impose such patterns of thought on activities in the future. Surely we know that inter-human work cannot be automated or rationalized? Surely we also know that intra-human creativity of the kind that opens up new knowledge (e.g. research) and disseminates existing knowledge (e.g. learning in order to prepare for insight) must be measured by other means than the old industrial units? 'Surely we know?' I ask, but it is the wrong question. 'When will we know these things?' is better. When will we identify good alternatives? It will certainly take a long time, unless we start un-learning past notions now. To do this we need thinkology, not technology.

THINKOLOGY VERSUS TECHNOLOGY

The thinkology of the future has had strict limitations imposed on it by the information- and media-driven society that followed industrialization. Here we really are confronted with high-speed change! One of the real problems of our present time is the horror of a break in the flow of media products or of chinks in their smooth surfaces. Only rarely do we encounter something that is hard to deal with in any way. This fear is driven by the possibility of consumer channel-swapping. The constant adaptation of the products can lead into a self-destructive spiral of tedium. The more easily digestible the content, the less able are the minds of the consumers to cope, and the more likely they are to zap. The expectations of the media programmers become self-fulfilling, in other words.

Digital TV has plenty of room for new channels and I think the moment has arrived for a new type. Called something like 'TTVF' ('Tempt the TV Fatigued'), it would be designed to soothe the nerves of viewers who can't stand watching any more. We have stopped watching, not because we don't like the programmes, but because the jerky pace drives us crazy. TTVF would have a commercial appeal, because it could proudly announce that its advertising breaks were quite out of the ordinary: 'No breaks unless the ad is at least an hour long – sell on TTVF! Get away from those indifferent, bitty trailers that do nothing except break people's engagement with the on-going programme.' Viewers would feel safe, because we'd be assured of continuity and proper intervals once in a while. The non-zappers would fall for a channel that prioritized thoughtfulness and a gentle change-over from one programme to the next.

THE INERTIA OF THOUGHT

For as long as the pace of progress stayed reasonable, our ability to recognize old patterns of thought was a good way of keeping in touch with our cultural inheritance. Having learnt to understand it, it was easy to pass it on to the next generation: 'All is as it has always been and always will be.' Then the pace of change accelerated, and became so fast that we were losing our ability to understand and retain traditions. Disorientation about the past has made it preferable to question old ways of thinking, rather than take the trouble of trying to interpret the present in terms of traditional precepts. This rejection of the past has not come easily. Not only are our bodies innately conservative, but so are our minds. Evolution has built in a distinct pacelag.

Our eyesight is a good example of evolutionary pacelag. We have good forward vision, but much less functional sideways vision. We only catch glimpses of what is passing us, but enough to get on with as long as we are moving at speeds suited to our own powers of movement. Fast travel on horseback had its problems, but our peripheral vision was still useful, particularly when head-turning helped with the focusing. It's quite different now, when driving has increased our speeds into the range where we can no longer act on a glimpse of a running child or the sudden appearance of another car. Work on technical aids is under way, but this does not change my main argument.

THE ABSENT PHOBIAS

It is interesting that the technical development of vehicles has moved at such a pace that we haven't had time to develop any speed phobias. In other respects evolution has been slow enough to allow phobias to develop in response to danger. Although extreme reactions, phobias about heights, empty or closed spaces, spiders, snakes and so on make sense. Presumably it is because moving at speed is such a recent phenomenon that no similar warning-system has had time to get genetically established.

Is there any way of making up for the lack of innate reflexes? Well, first of all we can consider the past and the collective common sense of old. It seems to work in many areas. One example of the shared caution that was still around when I grew up is the fact that everyone would have been horrified – though they had no fancy research results to go by – at the notion of feeding vegetarian creatures like cows with dead animal remains.

COMMON SENSE

Of course, we should scrutinize the old saws derived from our collective common sense with a great deal of scepticism, but they are not just an easy get-out from thinking. Ancient common sense is more likely to contain valid, lasting truths than more recent precepts. In the last fifty years, say, many patterns of thought have turned out to have no staying power at all. Mentally, you would do well to mark ideas from the 1950s and 1960s with a sticker saying 'Best before 2000'. That should stimulate you to look for alternatives, or else for good reasons to extend the idea's shelf-life.

Also, we should be alert to potential trouble whenever we come across groups of same-sex or same-age people. Warning bells should ring immediately. Get another mental sticker ready: 'Beware Uniformity!' By now, we know enough about the risks of uncritically accepting the dictates of small elites.

There are many examples of changes that only ten

or twenty years later have turned out to be desperately misguided. Consider the scourge of 'mad cow disease'. What is the central message of this gross mistake, one of the worst in our time? How did anyone get the idea of feeding herbivores, destined for human consumption, with ground-up cadavers?

There were scientific reasons, based on the cellular metabolism of cows. Forget these, and consider instead how groups of people actually arrived at the decision. They would have been small groups, rather like cell-clusters. I believe that totally counter-intuitive decisions, such as turning vegetarians into cannibals, are taken by groups that are either very small or very homogeneous or – worst of all – both. Then the group dynamics become dominated by a backslapping joviality that replaces discussion: 'Excellent idea, my dear chap!'

If small, homogeneous groups increase the likelihood of people losing touch with the accumulated common sense of their culture, it follows that large, hetero-geneous ones might be better at keeping valuable tradi-tions alive. Decisions should only be made by mixed groups consisting of men and women, old and young, scientists and humanists and so on.

THE PACELAG IN RESEARCH

Research is often seen as humanity's lamp illuminating the future, but it too can be affected by fast rates of change. Speeding up the research process itself threatens its very nature. How to make the new idea available for investigation? Who is best equipped to investigate? Who is to appoint the investigators, and in turn scrutinize their activities? If maintaining quality is the goal but the criteria are exclusively based on what has already been found out, the risk that innovation will suffer is great. It seems reasonable to worry when the obstacles to genuinely new projects are becoming almost insurmountable.

But then, how can beginners be credited with being right when the qualitatively excellent (at least by its own evaluation) establishment does not agree? By definition, the new idea lacks accepted merit, and the project might not even be based on well-tried methodology. It can be practically impossible to differentiate between genius and fantasist. Examining a new proposal is made even more difficult when its premises are unheard of within the current set of rules for deciding between projects.

It seems clear that discrimination of this sort will

become harder and harder, the faster the rate of growth in accumulated knowledge. Today, the sum of active researchers is greater than the sum of all researchers in the past. More and more societies insist that they are knowledge-based. One might say that the growth of research is exponential.

EXPONENTIAL FUNCTIONS

What is an exponential function? You must have seen graphs of exponential variables: the curves either start slow and flat and then suddenly take off upwards, or start on a plateau only to slope sharply downwards at some point. Upwards to heaven or downwards to hell. All these curves can be derived from similar mathematical descriptions.

I shall try to explain. If the very idea of mathematical explanations makes you feel sick, just leave the rest of this section alone and go on to the next heading. You might find it helpful to stay with me, though, and let me take you through the example of the spread of water lilies in a pond. No formulas: that's a promise! You will have learnt something new and generally useful in the end. In the twenty-first century it is impossible, or so it seems to me, to think aloud about time without mentioning the workings of exponential functions.

The number of water lilies in a pond is doubling each year. There are only a few plants in the first year, covering one-thirty-second of the pond's surface. People go swimming in the water and everyone admires the lilies. A year later one-sixteenth of the surface is covered, which still doesn't look much. By next year, it's different: one-eighth of the pond's surface is covered, increasing to a half over the next two years. And when only half the surface is visible, there is only one more year to go before it's all covered by lilies.

Draw a series of pictures, if you like. Don't you agree that the change is quite sudden? The basic rate of growth does not change. Each lily produces the same number of offspring each year, but because there are more parents every year, the total number of *new* plants increases from year to year.

This is one characteristic of exponential functions: the rate of change is stable. It doesn't matter whether it's an increase or a decrease. If the function is based on a rate of disappearance, again the process is slow initially and suddenly speeds up once the original number has declined below a certain level. Radioactive decay is a splendid example. When one half-life has passed, half the original number of atoms is left – and so on.

Exponential functions can be used to describe many processes in physics and biology. Biological examples include population changes before external checks have been applied, e.g. by limited food availability or environmental toxicity.

In technological contexts such as the rates of spread of electronic information systems, there are no obvious checks. The result is that the effects become over-whelming, since neither you nor I functions in exponential mode. On the contrary, we are very much attached to habits, i.e. steady states. In spite of this, we are becoming increasingly involved in exponentially changing processes, and these in turn tend to lead to profound alterations in our attitudes to time. Either we feel that time is running out of control, or else that the amount of change must have taken longer than it has – 'Not all that in one year, it's surely been two or three at least!'

It is more common to feel left behind than to feel part of extended time. Still, I have experienced the latter recently and maybe feel it now, as I write.

The knowledge-based society, the Internet and all

the things that take place in my mind are so extra-ordinary that something tells me these new things cannot have happened so quickly. It's like coming home after having been away on holiday for a long time. Large changes can lead to displacements in one's sense of time.

THE RACE FOR KNOWLEDGE

There are justifications for the view that chasing new knowledge is a bad thing on the whole. Could we decide that all new knowledge is evil by definition? What would be the consequences? I really don't know. Many, like myself, are innately curious and unable to stop looking for answers. Nor can we prevent one answer from generating more questions.

What should be done to manage the exponential growth in our accumulated knowledge before we collapse under the weight? True, many people can't be bothered even to look under the bed, but there is still enough research going on to turn most old truths inside out. The cloning of sheep-cells that created Dolly is a spectacular example, but only a start. Try to get your mind round the following statements:

- It is going to be possible to describe what life is, change it and create new life.

- It is going to be possible to disrupt the old relationship between sexuality and reproduction.

- It is going to be possible to blur the distinctions between life and death.

- It is going to be possible to reproduce learning in data processing devices.

Read them again a couple of times, aloud this time. 'It is going to be possible to...' means that on-going projects will deliver breakthroughs at a high rate in the near future. Then human concepts affecting all our everyday lives will start to change in fundamental ways.

Is it going to be possible to feel secure in the new world? Surely not the same kind of security, anyway. There are no rules rooted in social or mental structures to hang on to. The only saying I shall take with me from the 'old' world is a quote from Pippi Longstocking, heroine of the brilliant books for children: 'If you're very strong you have to be very kind as well.'

GOOD WILL, COMPETENCE AND COURAGE

There is no reason to assume that scientists are very kind. As a group, we are just like other people. Kindness is not a constant, though. People change as they interact with their environments, and of course the same is true of people who do research. Let us assume that most people will be likely to look for knowledge, and that most will realize how challenging the exponential growth in knowledge has become – also that more researchers will focus on human functions. Then what? Are these assumptions acceptable in the first place?

The methodological approach of research is not confined to small elites of highly trained scientists. Every day all kinds of people, including small children, try out new sets of connections between observations, or in other words, test new theories. A theory need not be particularly well thought-out or valid. It's enough that it applies for the moment, provided that the person finds it a useful tool to think with and the effects are reasonably harmless. Being generous about the ways people think seems a good idea, given that people like being allowed to think for themselves and that it is an enjoyable thing to do.

Of course, if you want to be *right*, you must have evidence to prove the excellence of your theory. People are not going to change their hard-earned views for less. Ideally, we should keep in mind Immanuel Kant's definitions of enlightenment:*

- Enlightenment is man's emergence from his self-imposed immaturity.

- Immaturity is the inability to use one's understanding without guidance from another.

- This immaturity is self-imposed when its cause lies not in lack of understanding, but in lack of resolve and courage to use it without guidance from another.

I would like to add some reflections about courage. I remember seeing a diagram in a newspaper, but have forgotten in what context: the two axes at right angles

---

* Immanuel Kant (1724–1804), highly influential German philosopher and the founder of Critical Philosophy. The quote comes from 'What is Enlightenment?', an essay in *Perpetual Peace and Other Essays*, translated by Ted Humphrey. Hackett Publishing Company 1983, p. 41.

were marked respectively *Good Will* and *Competence*. It is a good diagram to keep in mind. It has often helped keep my adrenalin levels under control in infuriating situations. The combination *Much Good Will / No Competence* is useless, and the reverse is frightening. A little of both qualities goes a long way and a lot of both means that things will work well.

However, a third quality is needed: courage. For instance, now that mankind faces a rapidly increasing strain on environmental resources, most of us are already sufficiently aware, concerned and competent. What we collectively lack is what Kant called 'resolve and courage'. Our mental diagram should have three axes: one for *Competence*, one for *Good Will* and one for *Courage*.

8

RHYTHMS AND NOT-RHYTHMS

Time is reflected not only in clock-time, intervals and speeds, but also in rhythms. There are many kinds of rhythmic functions: biological, mental, social and so on.

In physics, one of the ways to describe a rhythm is in terms of its frequency, that is, how many times one 'beat' occurs during one time unit. If the chosen time unit is one second and the beat (unit event) occurs once per second, the rhythm is said to have a frequency of 1 Hertz (Hz). The contraction/relaxation cycle in your heart takes about 1 second to complete, but exercise can speed it up to two or three times the normal frequency, i.e. to 2–3 Hz. If the unit event takes one hour to complete, i.e. happens once every 3,600 seconds, the frequency is about 0.3 thousandths of one Hertz or 0.3 milli-Hertz (mHz). There are smaller units still, of course. The point I want to emphasize is that it is almost always problematic to combine rhythms with very different frequencies.

My reason for talking about rhythms here is that I believe that innate differences in rhythmicity are much underrated as complicating factors, both between people and between people and their environment.

DIFFERENT RHYTHMS: THOUGHT AND SPEECH

There are marked differences between many internal (mental) and social rhythms. When you think, it is easy to move swiftly from one argument to the next in a long chain. If you felt like telling someone your thoughts – or had to, for some reason – the account might go on for ever, without even mentioning all the intermediate links in the chain. When once in a while you meet someone who thinks in the same rhythmic pattern as yourself, there is a mutual feeling of primitive joy at the ease of conversation. It's like finding the right partner on the dance-floor.

Of course your rhythms vary from time to time, but generally an individual seems to be tuned to a basic mental rhythm. This base-rhythm is probably one of the fundamental characteristics that differentiate one human being from another. The coincidence of individual rhythms surely influences person-to-person relationships.

I think there is a very special fascination about meeting someone whose rhythm matches your own. It was one of my colleagues who first drew my attention to 'thought-rhythms' and how they can either unite or separate people. With a good match, there's practically no end to the generosity that people show each other. On the other hand, if there is a mismatch the slightest thing grates disproportionately.

This kind of observation has led me to hypothesize that our minds can generate random rhythms that become fixed patterns of thought once they turn out to match another rhythm in a useful way. There would have to be brain mechanisms for generating these random rhythmic variations, with built-in timers controlling their frequencies.

Assume that this crude model contains a grain of truth. It would help to explain the sense of elation at meeting someone with the same rhythm, the sense that you're meeting another mind. It's a delight! The opposite is true too: it's useless trying to connect with someone whose rhythm is markedly different from your own. It's fascinating to speculate on how many people there must be on Earth with a mental rhythmicity to match your own. It stands to reason that you won't meet all that many. People are brought together by culture, religion and other interests, and perhaps especially by circumstances such as work and origin etc. – not because of their matching rhythms. The result is that we end up in tiny groups where perhaps we do not belong, because the most important thing about us, our thought-rhythm, is separating us from the rest.

Having thought about these things, the next step is to consider the possibility that each person has more than one base-rhythm, in the same way that we all have characteristic voice-frequencies. Intellectual and emotional states may have distinct rhythms. If these interlocking rhythms are important to the way we relate to one another, should we create group-identities based on rhythm selection? Probably not, but the risk of rhythm-based discrimination is so minute that it hardly seems worth warning against it. My only conclusion is that you might as well look for people whose rhythms match your own: meeting them is such fun.

INDIVIDUAL RHYTHMS

Ever thought about the word 'individual'? It means a person or thing that cannot be divided. One human being cannot be divided. Most of the workings of an individual mind are out of bounds for everybody else. I believe that most mental activity is inaccessible even to the person him- or herself, apart from some thoughts that are conscious and so available for control.

The shifts in my own thought-rhythms make me giggle now and then. The frequency is high most of the time, and sometimes so fast I can barely muster the energy to keep up. At the other extreme, there are times when my mind moves incredibly slowly. It can actually be quite relaxing. It happens that I wake up with just one thought in my head, and that thought then takes its time, recurring almost unchanged again and again. If it's a nice thought, it can be quite a pleasant state to be in. Physical exercise tends to turn me into a one-thought-at-a-time person. One thought, usually a rather dull one, lasts me right through a morning run.

It seems to me that high- and low-frequency thinking can occur simultaneously, sometimes cancelling out and at other times reinforcing each other. Thought patterns that partially cancel each other out might leave a single dominant notion behind. Such dominant thoughts are able to reset the whole mental machinery to their own rhythmic patterns. There are two models in my mind for such interacting rhythms. If a great wave comes along when you're swimming, all you can do is abandon yourself and float along with it. Also, many smaller waves with the right frequencies can build up into a great wave, which then overrides the remaining small ones.

SETTING A TREND, I

Some thoughts are seminal: they set trends in your mind. I'm going to tell you about some of mine, which have laid down long-term, fundamental frequencies for my thinking. The trend-setting thoughts impose both patterns and interpretations. New rhythms are more easily included if they reinforce these dominant concepts.

I was in primary school when I had my first realization of existential loneliness. The thought was triumphant and brand-new. It struck me like a flash of lightning: 'Nobody knows what I'm thinking! Nobody knows what I'm thinking, nobody knows what I'm thinking…' At least, I probably thought this again and again. I do remember keeping my lips tightly shut. Experiencing my thoughts as utterly mine, unless I decided to let them out, was tangible, and so very pleasing that the memory has stayed with me. My lips close tight just thinking about it.

Maybe my adult reaction sounds a bit silly, but it is a reminder that nobody can view my mental images, nor force my internal rhythms. This is true of everyone: if you get something of value out of this book, it's because you have read between the lines and adapted my ideas to make them your own. I would go further: what impresses you is likely to correspond to something already existing in your mind. All you need is the opportunity to let it arrive in your active consciousness. Your internal strings are already vibrating with rhythms that resonate with what you read. You might feel these to be 'new thoughts', but they were there all the time.

SETTING A TREND, II

I was a little older, maybe twelve or thirteen, when I had another trend-setting revelation about the role of the individual. A teacher, of handicrafts as it happened, said something that made me take note and ask myself: 'Is it right to say that?' Of course I should have asked her there and then, but I didn't dare. The situation, pedagogically speaking, couldn't have been worse: imagine a teacher saying something incomprehensible but being seen as so strict the pupil doesn't dare ask what they mean! Nonetheless, the effect was truly long-term. I still remember precisely what she said: 'It's so individually different.'

There I was, wondering if it made sense or nonsense. By now I've worked out that it's a remarkably wise saying. The phrase 'It's so individually different' has served me well over the years. It has helped me understand why people who have been in the same situation often describe it quite differently afterwards. Mental processes are so tightly linked to individual personalities that memories can quickly change into mere caricatures of the real situation. Establishing that both individuals were in the same sequence of events then relies on photographic evidence or matching accounts by independent witnesses. The reason could be that their mental oscillations were operating on such different frequencies that only selected events reinforced them in each case.

SETTING A TREND, III

Once, still quite young, I went to the theatre to see a 3-act play by Brecht.* Act 1: A woman prisoner in a concentration camp is wondering who might have denounced her. She finally decides that it must have been one of her neighbours. Act 2: The neighbours are on stage and we learn that they did not denounce the woman. By now they know that she's to be released after only half a year in the camp. They speculate about the reasons for this quick release and conclude that she must have agreed to spy on her neighbours. Act 3: The woman and her neighbours meet. They try hard to be friendly and trust each other, but cannot.

Once distrust has entered relationships it is hard to root it out. It is like a couple dancing to their private inner tunes rather than responding to the same beat. They are both unable and unwilling to adjust to a common rhythm.

---

* Bertolt Brecht (1898–1956). German poet, playwright and theatrical reformer whose theatre developed the drama as a social and ideological forum for leftist causes. The 3-act play is possibly *The Good Person of Szechwan*.

Lack of interest can also operate in this way. Among my colleagues there are individuals whom I admire a lot, with good reason – their intellect, creativity, diligence, integrity – but who cannot bear dealing with mathematical structures or symbols. If I'm presenting a complex argument and try my favourite way of simplifying it by using a formula or two, or a table giving some numbers, it is counterproductive with the non-mathematical members of my audience. The internal strings in the individuals may have been vibrating strongly and positively before, but then they stop instantly.

RHYTHM AND NOT-RHYTHM: SEVEN KINDS

Of course we can adjust to rhythms and tolerate not-rhythms, but there are limits to both. To round off this chapter, I have drawn up an unsystematic list of examples. It might stimulate you to think of your own.

1. Rhythms in nature

I belong to the majority of people who dislike ironing. But I forget about being bored if I take the laundry outside. At times I positively like doing the ironing in the garden. Obvious reasons might be that it smells nice or there are birds about or a mild breeze, but I believe there are deeper things afoot. The rhythms outside suit me; I find them restful. If I have to iron inside, I put on some music, although I haven't found anything quite right for lifting the tedium. My favourite music to iron by is one of the Brandenburg concertos. Why? I've no idea.

2. Water rhythms

I also belong to the large group of people who enjoy sitting watching the sea and just being. Maybe thinking. A stream, preferably dancing among rocks, is also good to watch. The sea seems to enchant us into feeling in tune with eternity – for eternity too is a kind of rhythm, a rhythm with a frequency of zero. Anywhere, gently streaming or undulating water entices us with its variety of rhythms – not to speak of the glittering dance of sunlight on the surface of a swift-running brook!

3. Our weak sense of rhythmicity

We human beings have some innate ability to sense rhythms and adapt to them, but other species are better at it. Much better, in cases such as the hummingbird. When a hummingbird is sucking nectar and the flower is waving in the wind, the bird keeps its position by swinging with the flower's movements. No human being could match such a feat. Our sense of rhythm is not well enough developed.

4. Rhythms in children and in adults

Some of the rhythms seen in children disappear or are even reversed in adults. Think of meal-times, for instance: children get noticeably more lively after meals and adults more drowsy.

5. Rhythms in traffic

Listening to traffic rhythms does nothing for anyone. Or at least no one I know, though such people may exist. But most people adapt to rhythms in traffic flow at different times of the day. Commuter traffic on local stretches of motorway is usually smooth-flowing and reasonable in the morning. The drivers are on their usual morning run and don't get over-excited about it.

6. Rhythmic intervals

It seems obvious that long-distance runners have a different internal rhythm from sprinters. It's not entirely about physiological stamina, sustaining either steady effort or brief explosions of power. It is also a matter of distances between intervals: the pause frequencies.

7. Meetings and their not-rhythms

I shall stop after this, the seventh example. Seven is a magic number, after all. This is a not-example of a not-rhythm. I have chosen meetings, because for me at least their main characteristic is that they follow no discernible rhythm. I find it almost impossible to participate in committee get-togethers: a bunch of people who discuss assorted topics in more or less (usually less) disciplined ways and then proceed to make decisions. I used to be able to make a pretence of joining in and then get on with writing letters or thinking new thoughts. I seem to have lost this ability now, and instead get caught by the pointless deliberations. It quickly becomes a waking nightmare.

This I must assume is a personal quirk, because committee-style proceedings are very well established mechanisms of management. Apparently, most people do not mind having their thought patterns locked into the not-rhythm of the back-and-forth across the table. This means, of course, that my own instinctive resistance matters less. But I'm curious: are some people actually attuned to the committee rhythms? Or could it be that they are relatively rhythm-insensitive and therefore more tolerant of not-rhythms?

9

THINKING FORWARD AND BACKWARD IN TIME

'Why?' is a very common question. Usually it can be answered from two different points of view, by looking either at the background or at the intention. This chapter is about the distinction between the two views. They can be expressed as respectively 'because of', i.e. looking back, and 'in order to', i.e. looking forward. That time can have direction is an important point.

We normally lead our lives looking ahead. Our expectations as well as our intentions direct us to do so. No one would even get up from a chair if he or she did not have another action in mind. Without being able to anticipate the future people would just stay put, like rocks. Having intentions, conscious or unconscious, is one of the distinctions between the living and the dead. Live organisms have innate and acquired intentions but dead matter has neither. For dead matter, the only causal connections lead back into the past. It is possible to track the changes leading to the present, but not – or at least only to a very limited extent – the effects of the present on the future.

It is especially important to be forward-looking in the context of education. Oddly enough, we often focus on preparatory knowledge, i.e. past knowledge. This is undeniably useful, but deals only with what has been. The crucial words to keep in mind are those which point ahead: expectation, hope, intention.

THE TELEOLOGICAL AND THE MECHANISTIC

Let us start with the 'backward thinking' or mechanistic approach, which describes how chains of events originate 'because of'. Science is made to carry the responsibility for the clockwork perspective on the universe, believed in by large numbers of people. In clockworks everything can be referred back to past events, already built into the mechanism.

First of all, it is unfair to insist that this is what science states. True enough, the physical sciences are interested in the natural laws that govern observable phenomena, but these laws are human descriptions of how nature operates. It is not right to blame science for the wider use of this kind of analysis.

Physics has in fact got nothing to say about the biological capacities of human beings. When I reflect on human behaviours, for instance, on human memory, it is not as a physicist – I am not 'a scientist', but a person who happens to be in that line of work. It is from the 'person-perspective' that I speculate on the consequences of different theories about memory.

The theories come as two main variants. One type says that everything held in our memory-stores remains there unchanged, and the other that everything is continuously restructured. Those who believe in the first kind of theory will find it hard to escape the conclusion that people are always governed by their past, whether they are conscious of it or not. People are chained to their previous experiences, as they are fundamentally unable either to re-evaluate or erase them. It follows that our lives can be understood in mechanistic terms. Most things happen 'because of' and not because we intended something new.

The other extreme form of memory theory states that the past is restructured all the time. Memories are not just records constantly ready to be compared with what is current, but also subjected to constant change.

The two variants are used in parallel, but in practical contexts they can lead to markedly different conclusions. The psychologist's testimony based on interpretation of a child's drawings and stories will depend on whether he or she believes events are turned into indelible mental images, or vague elements in a changing narrative. In a court of law, the judge and jury should establish which is the psychologist's theory of memory, before examining the expert evidence.

THINKING BACKWARDS – FOR BETTER OR FOR WORSE

Our experiences shape us to a large extent. Our knowledge, associations and emotions are tied to our experiences. It is impossible to imagine a human being who does not look back at his or her past; even a newborn has many experiences stored away. Later in life our past stays with us and greatly influences who we are and what we do, in small matters and large.

Think of two people jumping from stone to stone along the edge of the sea, and compare them. They chose different routes, and jump differently too. With each new jump our mind unconsciously, and with lightning speed, checks through the characteristics and outcomes of a number of previous stones, speeds and foot placements. Most options are rejected because experience tells us that they are less likely to be successful. It is the individual's combination of reality-check and experience that results in the choices of stones and jumps. There are also individual differences in the process of choice. Just as well. Individual development would surely grind to a halt without that innate drive to test new possibilities. I must restate – it cannot be said too often – that learning is dependent on variation, not repetition. But the stored past is also necessary as a basis for variations. It provides an awareness of the possibility of new ideas, and a stock of memories for comparison.

Holding on to the past is pointless if the situation is utterly new: then old experiences are no longer useful. A small child without experience of weak ice-cover on lakes will walk out on thin ice without a qualm. Schools without experience of the Internet and its use in learning will unthinkingly assume that the old teacher-pupil relationship is still the only valid one.

LINEAR VERSUS CIRCULAR THINKING

Our culture is pretty exclusively based on linear thinking. In a culture with this characteristic, it is obviously as important to work out the consequences of thinking forward as those of thinking back. Would it be right to assume that in cultures based on circular thinking, this directional analysis might not be as crucial? Let us examine two classical traditions of thought a little more closely: the Chinese and the Greek.

Our culture is strongly influenced by classical Greek culture. It was part of its world-view to regard everything as composed of mixtures of four elements: earth, water, air and fire. These elements conveyed proportions of four properties: coldness, wetness, dryness and heat. In the human body the corresponding elementary ingredients were body fluids: black gall, white phlegm, red blood and yellow gall. The corresponding human traits are: melancholic, phlegmatic, sanguine and choleric. The model also emphasized that the four elements and their qualities were independent of each other and in no sense ordered hierarchically.

The classical Chinese view was that there were five

elements, excluding air: metal and wood, in addition to earth, water and fire, in that order. In this model the elements were interdependent and had to be aligned in sequence, but also preferably around the circumference of a circle. This meant that metal (1) bites into wood (2); wood bites into earth (3); earth bites into water (4); water bites into fire (5) – or in a similar sequence of simple acts: metal cuts wood, a wooden rake cuts into earth, earth soaks up water and water puts out fire. Then, the perfect fit: fire (5) bites metal (1), for fire melts metals. The circle is closed.

It seems reasonable that the opposition between the two modes of thought – Chinese versus Greek, circular versus linear – can be traced to our own perceptions of time. Do we see it as circular or linear?

Our linear culture has shifted the direction along the axis of time from the forward-looking perspectives of the ancient Greeks to our contemporary stress on the sequence of cause coming before effect. Imagine a Greek child had found an acorn and asked: 'Why is an acorn lying there?' It would have been told: 'Because the acorn is going to grow into an oak.' This is an example of teleological thought, which means that the mental model is one of explanations directed towards particular ends. Nowadays, the answer would be different. The child would be told: 'Look around, dear – there's an oak there. The oak has dropped the acorn.' In our mechanistic culture, cause must precede effect.

Only a small number of people have got the measure of technology, our modern driving force. Building a clock? It is planned *because* it is meant to show the time. The mechanic somehow builds his intention into the clock. Afterwards, he can apply his scientific model and explain that the clock works because of this or that design feature. This does not change the primary intention, namely that the whole process was started from a 'because of' perspective.

Consider this: maybe we could learn a lot by looking for intentions and not just causes. All new knowledge, from the starting moments of the universe to the most minute details of a microcosm, astonishes us by showing how functional, how miraculously intentional, the whole structure of nature seems to be. Yet we stubbornly refuse to look for structures of acquiring knowledge that are based on intention. Are we being reasonable? Should we not allow ourselves to search for a way to be economical with thought? And, as a consequence, stand up in defence of thoughts that suit the problem and feel right in the context. Maybe consider the possibility that some people are drawn to research by a wish to follow up 'in order to' rather then go looking for 'because of'. Maybe it would suit women?

FORWARD THINKING – VISIONS

Forward thinking tends to be about visions, that is inner images of how things might be and perhaps also how to set about making the images come true. I remember going round a major exhibition of Leonardo da Vinci's work, most of the time deep in thought. It was awesome to realize how many new things da Vinci must have envisioned. Arguably his greatest, most important, gift was this capacity for crowding his mind with almost tangible images. Naturally, his outstanding ability to draw and paint made him well known to his contemporaries and an immortal artist in the eyes of later generations. The intensity, variety and realism of his inner visions are represented in da Vinci's rich and apparently inexhaustible flow of work. To me at least, this mental inventiveness is his most remarkable characteristic.

Even though we may not be able to express ourselves by drawing, we all have inner images. We 'see' them with varying clarity and use them more or less deliberately. Generally, these visions control much of what we do. The more seemingly real and the more charged with emotion, the more they direct our actions. It is an unhappy fact that most people and most social systems rarely experience a new idea, however good, as real in that sense. It takes time to establish it enough to join the directive visions.

On the other hand, the existing alternatives to the new idea will already have come to be seen as real, and hence as obstacles to innovation. Worse still, established ways are often infused with emotion. Even good, constructive visions lack realism and emotional charge and so tend to lose out to imaginary problems. Lots of ideas simply get strangled at birth.

FORECASTING AND BACKCASTING

One way of facilitating the new is to turn time round by 'backcasting'. Try to imagine what things will look like five years from now, then place yourself mentally at that point in time and look back into the past. Try to identify the reasons for the future present being the way it is. Outline the timetable and pinpoint stages, deadlines, before-and-after situations and so on. My own experience as a manager of project work, both in research labs and elsewhere, tells me that many people find it remarkably difficult to reconstruct their speculative goals into concrete proposals. By turning the whole process round and then looking back at it, planning is often made easier for co-workers.

True as it seems to be, this observation baffled me at first. I now believe the reason lies in the way we use our memory to explain how we came to be where we are now. The mind is used to analysing past events. No surprise, then, if the method works for future events, as long as the imaginary move forward in time allows the same direction for the analysis.

One of the advantages of 'backcasts' is that obstacles look less troubling than they would in forecasts. Looking back is what we make our memories do, but in a backcast there are no actual problematic memories. On the contrary, some of our memories might diminish the importance of obstacles. Maybe it is a good way of thinking realistically about the future: cancelling out the problem-enhancement of forecasting by the problem-reduction of backcasting.

Other advantages come to mind. One is inspired by a classical philosophical argument, based on the strategies for occupying an enemy city. The debate focused on ways of negotiating the city wall. Would they use ladders and try to take the city by storm – or maybe first besiege it and starve the inhabitants? Maybe a rain of arrows? Fire? The attackers were absorbed by such technical problems until someone demanded to know the answer to a new question: 'Assume that we have taken the city – what do we want it for?' When they concentrate on the answer to that, they may well decide against attacking at all. Alternatively, it will make it easier to adapt the method of attack to the overall campaign strategy. Of course, my own interest in back-casting is related to its use in solving civil problems, not military ones, but it remains an important problem-solving technique.

SACRIFICES TO UNREALITY

My arguments so far assume that we let no fantasy into our fore- and backcasting exercises but keep our eyes strictly on a real ball in a realistic court. Once unreality creeps in, it doesn't matter which direction you look in.

Maybe it was a sign of this that I sensed unreality in so many of the displays in Expo 98, the world exhibition held in Lisbon. I was surrounded by product logos rather than by real products. The whole site seemed to be dominated by organizations showing off fantasies, abstractions and symbols. Is this where we have got to?

Another but similar line of thought came to me when I compared our common inheritances of environmental resources in land, water and air, and of tele-culture manifestations such as broadcasting, electronic financial systems and entertainment industries. The media affect us every day and are insidiously diverting us from thinking about the world around us, from the atomic level onwards. If the media world has become more dominant to people than their environment, then the ranges of the electromagnetic spectrum which sustain tele-culture must be one of nature's most precious resources. No one notices, because these waveforms are invisible. They cannot be seen or traded in, and their distribution is decided quietly, in a series of international agreements. The most ordinary people might learn about these is in tiny newspaper items.

In the famous book by Antoine de Saint-Exupéry there is an exchange between the Little Prince and the fox that expresses very well the link between the past and the future:*

---

* Antoine-Marie-Roger de Saint-Exupéry (1900–1944), French pilot, poet and writer. *Le Petit Prince* (*The Little Prince*) was written in 1943 and is widely admired both by children and adults.

'The rose is important to you because you've been caring for it for such a long time.'

'It's because I've been caring for it for a long time...' the Little Prince said.

'It's a simple truth, but people have forgotten it,' the fox told him. 'Don't you forget, though. If you have tamed an animal or something, really got fond of it, then you're responsible for it always. You're responsible for your rose...'

'I'm responsible for my rose...' the Little Prince repeated so he would remember.

When you have become fond of – 'tamed' – somebody or something, then that will give you your strongest anchoring point in life, regardless of whether the object of your affections is real or fantasy. Fondness still gives meaning to both your past and your future.

10

Is there a Poodle Shortage?

Your idea of time is based on a few fundamental perceptions. Some of these may have been changed by reading this book. If so, remember that it's important that you let the new ideas change your view of the world around you. To help you and maybe also to make you laugh a little at yourself, but kindly, start by laughing at other people's notions and perceptions. Once more, we'll mentally travel to Greece. We'll focus on the number ten.

This is the tenth and last chapter in my book. This is not by chance: I had decided that it was what I wanted. Not just because we have ten fingers and toes and a decimal system based on the number ten, but because I wanted to have a good reason to describe the role of number ten in the Greek mysticism about numbers.

Integers were regarded as valuable at the time, especially the first ones in the series:

1 corresponded to points
2 corresponded to lines
3 corresponded to surfaces
4 corresponded to volumes

By then the dimensions of space were exhausted, and the dimension of time was still poorly understood. So, since 1, 2, 3 and 4 were sacred numbers, the sum of these numbers must be more sacred still: 1 + 2 + 3 + 4 = 10!

The number ten became dominant in the patterns of thought in ancient Greece. When they managed to count only nine separate celestial phenomena, by briskly mixing Sun, Moon, the planets and some of the stars, they just *knew* something was missing from the list. This was tricky, and they were reduced to inventing the counter-Earth. It was situated behind the Sun: when, as they thought, the Sun circled the Earth, the counter-Earth moved simultaneously and stayed out of sight.

This was a truth that could be neither contradicted nor investigated. Such 'truths' are dangerous. Something is because it must be. Note that no one thought about the counter-Earth in terms of an invention. It existed, because it had to exist. This is not too dissimilar to our current conviction that we are all short of time. It's not particularly intelligent to insist that we're short of time because we're short of time, but we still believe it.

PERCEPTIONS OF TIME

Everybody has their perceptions of how things must be. If necessary, we fix things so that they conform to our perceptions. This book is full of my thoughts because that's how I *want* it. Is this as comical as the Greeks' reconstruction of reality? So, I may come across as funny, but I don't mind. There's far too little fun and laughter about these days.

'Time to laugh' is one item in a list of eleven fundamental activities that people should find time for, drawn up by Guido Schwarz.* Schwarz classified

---

* Guido Schwarz is probably the Swiss initiator of 'Qualitative Motivational Research' (www.motivforschung.cc). The title of the work containing the list in the text translates as *Fundamental Needs and the Time They Take*.

essential functions among his 'needs', and also activities necessary for survival. Check out the list below – do you agree?

- Sleep
- Drink
- Eat
- Excrete
- Protect yourself against excessive temperature change
- Engage with others socially
- Laugh
- Engage with others sexually
- Weep
- Experience ecstasy
- Think

PARROTS, CHAMELEONS AND POODLES

It demands a certain effort to rid yourself of habits which have become associated with fixed time-slots. Other habits that are difficult to uproot may have become established through the feeling that you're pressurised by lack of time or that your existence lacks meaning. Let me introduce you to an internal menagerie that can be put to good use: parrots, chameleons and poodles.

The parrot stands for the mimicry and imitation we often fall for. Most of these affectations are reassuring because we recognize them. With them come a constantly packed diary, fragmented time, ringing phones and a sense of increasing inability to cope.

Next, consider some chameleons. There might be something of the parrot about them – so busy! – but they can change colour and shape. In the world of the chameleons it is permissible to talk of undisturbed time and set-up time, as long as you still deliver the same amount of completed tasks. This is an improvement, and might make you feel more at ease.

But poodles are better. They represent getting to 'the core of the poodle'.* What I mean is to find what lies behind the obvious, get past the façade and reach the innermost essence. Poodles are rare. You may find it hard to get round to their way of thinking. The difficulty

---

* In Goethe's *Faust* (begun in 1773), Mephistopheles adopts the guise of a poodle to approach the master on his walk. Often quoted: 'So that was the poodle's core' ('das also was der Kern des Pudels') and, 'Don't growl, poodle.'

is to free yourself to think afresh. Sometimes I think 'time poodles' are particularly rare. It might be that innovative Thinking about Time Takes More Time Than Most.

If you get a clever idea and find 'the core of a poodle', you may still not dare speak about the idea, or if you do, people might not allow you to go ahead and try to make it reality. The most problematic obstacles are normally based on fear of the new. Others are set up because people fail to understand. The new thought does not fit in with the old view of the world, and the new idea seems hard to grasp and somehow *wrong*.

Often, people distancing themselves from innovation refer to the Great Risk: 'What would happen if everybody did this?' Just say, as you should: 'Fine, just fine, and besides everybody won't!' The possibility that poodles might dominate parrots and chameleons is minute. It is the poodles who are near extinction and rarely get time to develop properly, from birth onwards. One of the worst threats to them is the widespread pessimism about our future. What's the point of doing new things when the future is coming to an end?

PARROTS AND INSTANT MIMICRY

The West seems to have a monopoly on pessimism about the future, but recognizes no responsibility to prevent what is supposed to happen. The black vision may be founded on solid facts, but it always misses the most essential counter-argument: that which relies on human nature. There is no recognition of the possibility of human action causing sufficient change for the better, either in the short or the long run.

The following words were written by Dietrich Bonhoeffer* while he was incarcerated in a concentration camp and waiting to be executed. The quote shows that he was indeed able to see 'the core of the poodle' and hang on to his perception. Can you?

> In essence, optimism is not an evaluation of
> a given situation but a life-force. A force that

---

* Dietrich Bonhoeffer (1906–1945). German priest and theologian. Executed in the Flossenbürg concentration camp in 1945 for his role in the resistance against Hitler. The quote is probably from Bonhoeffer's last letter to George Bell in England and (if so) is to be found in *Letters and Papers from Prison*. Enlarged Edition, Eberhard Bethge (ed.), Macmillan, 1971.

enables you to hope when others have resigned and gives you strength to endure disappointments. A force that will not let go of the future and leave it in hands of the pessimists, but annexes it in the name of hope.

The extent to which your own pessimism about the future has affected you – should you be a pessimist, that is – probably depends in part on your job. A teacher who no longer hopes for the future should probably think seriously about resigning. Schools are institutions in which the work is directed towards the future: they should have no place for someone who is pessimistic about it. Spreading dismay and despondency among children would surely count as grounds for dismissal?

It is a widely held belief that 'It is undignified to despair'. I would go further and argue that it is not human nature to give up, to despair. It's not part of the human constitution. Imagine someone who cannot swim but who ends up in deep water. It's extremely unusual, even as a mental experiment, for that person to tell himself (or herself, as the case may be): 'OK, I can't swim, that's it. I'll sink.' Instead he will splash along as best he can. It might work, or on the other hand it might not. No one gives up just because a good outcome looks unlikely.

The future is not like a roadblock into which we will all crash simultaneously. It is no precipice over which we will all tumble one day. Instead we will continue to do what human beings have always done: keep trying.

If you share this insight, foster it. It's a primary condition for thinking constructively about time. Don't worry if you meet with aggression. It's not easy for people to cope with a poodle when they're used to parrots.

THINGS I KEEP WONDERING ABOUT

I wrote in my introductory notes that 'I have been thinking about time … on and off for the last twenty years.' What do I see if I look twenty years ahead instead? What is it that I would like to know? Every time I've written about a particular line of thought many more, mostly questioning, have come to mind. One of the most challenging questions is: what will be the amount of change that can occur during a human lifespan? The time between the present and 2020 is less than a human life span, but I am more interested in the trend.

Man is man's measure and the duration of human life is one kind of measure of time. Another measure, this time of a rate, is the pace of change that human beings can accept. Of course, the relationship between the change and the effect is not always predictable. Take a simple example: if the number of lunch options you can choose between increases from 10 to 1,000, the effect on you personally is minor. You are the measure: it's not really possible to eat lunch more than once a day. On the other hand, the growth in choice means that you have a greater opportunity to distinguish yourself from people around you. This will affect your interaction with others, for better and for worse.

Still more important is how you internalize social interaction – or so I believe. What you retain and reject can change the way you understand the world. Up to a point, anyway. Can one generation really take any amount of change in its conditions? Maybe the rate of ageing will increase as a natural response? Could it be that the human lifespan will shorten in inverse proportion to the increasing rate of change?

There is nothing to say that the rate of development will stay constant, so that, for instance, two changes in 2000 will be followed by two more in 2001 and so on. I cannot be sure that there will be just forty changes on record by 2020. Or maybe the rate of change will double each year – but if you tried to calculate what that would mean by 2020, you'd arrive at an incredibly large number.

I do not believe that the number of major changes can be counted in this or any other straightforward way. At the same time, I hear the ticking of the recorders running at an exponentially increasing pace. So I wonder. Because I do, I cannot help believing that it is important that many more people should also wonder about our relationship to time.